COLOR TV SERVICING

Other Books by the Author:

Television Servicing, 3rd Edition—Prentice-Hall, Inc.

Fundamentals of Television—Hayden Publishing, Inc.

Buchsbaum's Complete Handbook of Practical Electronic Reference Data—Prentice-Hall, Inc.

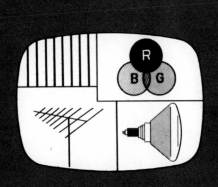

COLOR TV
SERVICING
THIRD EDITION

Walter H. Buchsbaum

Prentice-Hall, Inc.   Englewood Cliffs, N. J.

Prentice-Hall International, Inc., *London*
Prentice-Hall of Australia, Pty. Ltd., *Sydney*
Prentice-Hall of Canada, Ltd., *Toronto*
Prentice-Hall of India Private Ltd., *New Delhi*
Prentice-Hall of Japan, Inc., *Tokyo*

Library of Congress Cataloging in Publication Data

Buchsbaum, Walter H
    Color TV servicing.

    1. Color television--Repairing. I. Title.
TK6670.B8 1975   621.3888'7   74-11350
ISBN 0-13-152397-X

PRINTED IN THE UNITED STATES OF AMERICA

ABOUT
THE THIRD EDITION

While the fundamentals of color TV have not changed over the years, the actual color TV receivers—and the methods required to service them—have changed considerably. Since the first edition of this book was published, color TV progressed from a fascinating curiosity to an essential household appliance. The old 30-tube consoles have been replaced by transistorized table models and eventually by color sets using integrated circuit chips and featuring much improved color picture tubes. The third edition increases the practical value of Color TV Servicing because it includes all the latest circuits and deals with new components such as transistors in the same practical manner that has made the first two editions so popular.

Readers of the second edition especially appreciated the organization of the chapters of this book and we have therefore continued it. The first three chapters cover the fundamentals of color TV as they relate to the most recent developments. We again urge the reader not to proceed beyond Chapter 3 until he has a clear understanding of these fundamentals. Chapters 4 through 11 are devoted to those TV receiver circuits that apply especially to color sets, omitting the sound section and other circuits which are the same in monochrome receivers. Each receiver section is first discussed in functional block diagram form, covering the input and output signals and the circuit functions. Then typical tube, transistor and IC circuits are described.

Installation, adjustment and alignment are such important parts of color TV servicing that Chapters 12 through 15 concentrate on these topics in a practical, step-by-step approach.

The last four chapters will increase your skill in troubleshooting. Our readers have found this portion of Color TV Servicing so valuable as a practical reference and troubleshooting guide that, although it now reflects the latest techniques with new types of equipment, we have retained the same unique format. Chapter 16 will help you locate defects in the monochrome operation of a color set. Chapter 17 covers defects which cause lack of colors and Chapter 18 will help you uncover the reason for incorrect colors. Symptoms are broken down further according to the primary color affected, such as "insufficient blue" etc. Chapter 19 covers all those defects which defy classification, whether they are due to interference, intermittents or even faulty repair.

A practical book on color TV receivers could not be written without the cooperation of receiver manufacturers, and I want to thank the service and public relations managers of all the TV manufacturers whose credit lines appear in this book. I also want to acknowledge the diligent and clear manuscript typing by Mrs. Inge Seymour and the color photos by Jim Johnston. As in past editions, my wife deserves my thanks for her encouragement and active participation in the development of this book.

W. H. Buchsbaum

CONTENTS

1

INTRODUCTION TO COLOR TELEVISION

This book is designed to give the reader knowledge of the principles of color television and to teach servicing of color TV receivers. Understanding black and white television is a prerequisite to any course in color TV, just as a knowledge of radio is necessary to understand either branch of television.

The reader should know the principles of radio, TV receiver circuitry, and have some knowledge of television servicing before starting this book; an understanding of these subjects is taken for granted and well-known principles are not explained again. The field of color television contains so many new concepts and is so complex that this entire book is devoted to stressing and discussing the salient new features of color rather than rehashing the familiar principles of monochrome TV.

Today's color TV receivers make use of the most advanced state of the art in electronics. This means that, in addition to vacuum tubes, the latest color TV receivers employ transistors and integrated circuits. Both of these solid state devices are also found in increasingly more frequent applications in monochrome TV receivers, particularly in the smaller, portable sets. Integrated circuits are coming into ever increasing use in high fidelity audio equipment as well as in color TV receivers and the special methods of troubleshooting equipment using integrated circuits should be at the finger tips of every technician, even before he enters color TV. Those of our readers who are not familiar with transistor circuits should refer to the very excellent text, *Transistor Circuits and Applications*, by L. Cowles, published by Prentice-Hall, and those who want to learn more about integrated circuits are referred to, *Handbook of Integrated Circuits*, by H. Thomas, published by Prentice-Hall.

Manufacturers have replaced individual functional sections of the color receivers with integrated circuits, while special power transistor circuits replace vacuum tubes. This process will continue until, eventually, all TV sets use integrated circuits for the low power functions and transistors for all high power functions. It is impossible to anticipate the detailed circuits that will be used but this text will provide the reader with a clear description of the functional requirements of each circuit so that, whether it is tubes, transistors, or I.C.'s the reader will be able to handle any defects.

Having forewarned the reader about the prerequisite knowledge he needs, we shall now illustrate the major difference between the familiar monochrome and the new color television system.

Comparing Monochrome with Color TV

Television broadcasting spans our country, with more than 600 different transmitting stations interconnected by various types of networks. Each of these stations can transmit either color or monochrome signals when the proper equipment is installed. A total of 82 channels are assigned among these stations and no channel assignment need be changed to allow color transmissions. In other words, any station that transmits monochrome TV signals can also transmit color telecasts on the same channel, using the same transmitter, and serving the same community.

Similarly, all present black and white receivers can receive color telecasts, from the same stations as monochrome transmissions. Since the black and white receiver can reproduce only black and white pictures, the color telecast will appear as an ordinary monochrome transmission. On color TV receivers, the color transmissions appear in full color and the black and white signals produce the same kind of picture as on monochrome receivers. This feature of being able to receive color telecasts in black and white and monochrome signals in monochrome on color sets is called *compatibility*. The NTSC color TV system is compatible with the older monochrome standards; this is one of the major features which made the rapid and painless introduction of color

TV possible. Because of this compatibility, a sponsor paying for a color program can be assured not only of the audience having color sets, but also of the much larger monochrome audience. Lack of this compatibility had previously doomed another system of color TV.

Since the service technician will be concerned chiefly with color TV receivers, the difference between the monochrome set and its color counterpart is of great interest.

First, consider the receiver sections that will remain identical for both sets. A simplified block diagram of a monochrome receiver is shown in Figure 1.1. The tuner amplifies signals received from the antenna, selects the desired channel, and changes the frequency from the transmitted RF to the IF for more amplification in the following IF section. Superheterodyne action, local oscillator, mixer, IF amplifiers, and second detector will be familiar terms to radio and TV technicians and need no further explanation at this point. In the block diagram of Figure 1.1 an intercarrier type circuit is shown because this is the universally used system in all modern TV models and is also used for color sets. The FM sound section and audio amplifier stages are also universal and familiar circuits.

From the second detector the composite video signal is amplified and applied to the picture tube. A portion of this signal goes to the synch separator section, where horizontal and vertical synch pulses are separated from the video signal and fed to their respective scanning sections. The vertical sweep circuit supplies the sawtooth current for vertical deflection at the picture tube and the horizontal section generates the high voltage for the second anode as well as the horizontal sweep signal. A B-plus and filament voltage supply is part of the receiver and furnishes the necessary power to all sections. Transistor TV sets need no filament supply and often can operate from batteries.

While actual circuitry may vary between individual manufacturers and their various TV models, the basic sections outlined in Figure 1.1 are essential for every monochrome TV set. We have previously stated that color TV receivers can also receive and reproduce monochrome pictures; this indicates that all of the functions required for monochrome will also be found in the color set. A look at the block diagram of Figure 1.2 bears this out. All of the sections listed in the block diagram of a monochrome set are also present in the color receiver, plus some new ones.

Again the tuner selects the desired channel, converts the RF into the IF and the IF section amplifies the signal until it is strong enough to be detected. Although the block diagram does not show it, the tuner and IF characteristics for the color set are somewhat more complex than for monochrome reception. Bandpass, gain, and phase delay characteristics must be maintained to much closer tolerances than in black and white sets. The intercarrier sound section, FM detector, and audio circuits, however, are practically identical to those in monochrome sets. While a simple video amplifier sufficed in monochrome, a more elaborate *brightness* channel is used in the color set, although during monochrome transmissions the regular video signal alone passes through this amplifier. For color reception the operation of that stage is somewhat different.

Returning to the similar features, we notice that the vertical and horizontal sweep sections perform the same functions in both types of set. The synch separator section again separates the synch signals from the video, but in addition to furnishing horizontal and vertical pulses, it also supplies the color synch signal to the color synch section.

A completely new and different portion of the receiver is the decoder and matrixing section which is used only for color telecasts. In these two sections the actual color signals are decoded from the composite trans-

Figure 1.1—MONOCHROME RECEIVER BLOCK DIAGRAM

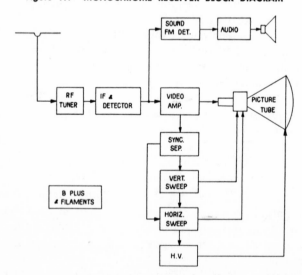

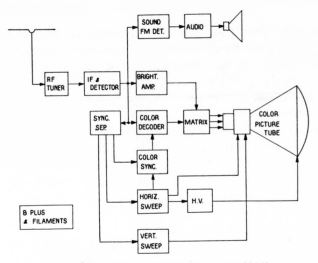

Figure 1.2—COLOR RECEIVER BLOCK DIAGRAM

mitted video signal and, after combining the brightness signal with the color information, three separate video signals are produced. Each of these signals represents one of the primary colors, red, green, and blue, which are used by the color picture tube. So new and different are the circuits used in the decoder and matrixing section that several chapters of this book are devoted to an explanation of them, with additional information on adjusting and repairing the decoder and matrix sections. The color synch section contains many novel features and is also discussed in great detail.

Although the sweep and high voltage sections perform the same functions for both types of receiver, in the color set the deflection linearity must be much better and substantially more HV is required by the color picture tube. In later chapters we shall see that these circuits are not basically new, but are more elaborate and slightly more complex than in monochrome sets. One radically new and different part of every color receiver is naturally the color picture tube. An entire chapter deals with the various color tubes now in use and additional space is devoted to adjusting and troubleshooting the color tube and its associated circuits. It should be understood that although the color tube is quite different, it is still a cathode ray tube and still uses a phosphor type screen, electron beams and magnetic deflection.

Having seen the functional difference between monochrome and color receivers, the outward appearance of the color set will hardly surprise the reader. It is larger, heavier, and more expensive than its monochrome counterpart. Figure 1.3 shows a monochrome and a color TV set side by side to illustrate the difference in size and appearance. The power supply, heat, and space requirements of the color sets are inevitably greater and most of the earlier receivers were manufactured only in console models.

Considering the increased size, complexity and cost of a color set, the customer will not be surprised to learn that servicing and installing a color set consequently also costs more than its simpler colorless cousin. The reason for this relatively high cost lies in the expense involved in training qualified personnel, buying additional, special color test equipment and—as the major factor—the anticipation of numerous "nuisance" calls. These plagued early monochrome TV too and were due largely to the customers' ignorance of even the simple front-panel adjustments. In color TV the additional factor of individual color preferences tends to complicate matters.

The complexities of color TV servicing and all the problems connected with color TV merely stress the great need for more service technicians. Without trained personnel color TV is simply not feasible. Since the personnel problem is so pressing, anyone looking for a good future will find color TV a fast-growing field of opportunities for the right man. Because color TV is a relatively new medium, the familiar cry for "years of experience" will not hamper the capable newcomer since there simply are not enough technicians with such a past in color TV. It is a free field, wide open, and needs trained help. The major accomplishment required of the technician is a thorough understanding of the principles of color TV. With this understanding he will gain experience and become an even more skilled and valuable man. Without the fundamental knowledge of how color TV works, even the cleverest serviceman can be nothing but a screwdriver mechanic.

We have gone to some length to impress the reader with the need for understanding fundamentals, because in the case of color TV the fundamentals often seem too theoretical and unrelated to the simple business of fixing receivers. Actually, the very complexity of the color receiver makes it necessary to understand thoroughly the operation of every portion, since it is otherwise a hopeless task to find a particular defect.

There may be a temptation for the reader to skip over the next two chapters and get right into the description of the color TV system. If Chapter 4 is studied seriously, it will become apparent that some previous knowledge is needed. That knowledge will be found in Chapters 2

Figure 1.3—MONOCHROME (left) AND COLOR SET

R.C.A.

and 3. We earnestly urge the reader to proceed from chapter to chapter, rereading parts that are not clear the first time and so gain a good understanding of the basic concepts, the fundamental rules, and the guiding principles upon which color television know-how must rest.

2

PRINCIPLES
OF COLORIMETRY

The Visible Spectrum

The human eye is a very unique organ. One of its features is the ability to see a relatively wide spectrum of electro-magnetic radiation. This is the same radiation phenomenon that is known as radio waves and is used in radar, diathermy, and other electronic applications. While amateur radio frequency bands range up to about 300 mHz or 1 meter, some of the precision radar systems use wavelengths as short as one centimeter. One meter is slightly more than three feet long and is divided into 100 centimeters or 1000 millimeters. The wave lengths the human eye can receive are in the order of less than a millionth of a meter, also called a micron. Actually the visible spectrum for most people extends from 0.4 to 0.7 microns. A certain color at 0.5 microns therefore corresponds to 600 mega-megaherz of radiation. Scientists use a still smaller term, the millimicron—one thousandth of a micron. Another expression commonly used in color or illumination engineering is the Angstrom unit, which is one ten-thousandth of a micron.

For the purposes of the color television student the spectrum chart of Figure 2.1 is accurate enough, using millimicrons as units for wavelength. Note that the "low" frequency end of the visible spectrum contains the red colors and borders on the infra-red and heat radiations. The "high" frequency colors are blue and violet, bordering on the ultra-violet type radiation. Some distortion is contained in Figure 2.1 in order to permit clearer color presentation, because in nature some colors take up a much narrower band than in the illustration. The colors shown in Figure 2.1 are called the pure or spectral colors, because each is made up of only one frequency. In nature these colors are seen very rarely and we usually observe only mixtures of these spectral colors. It is possible to see spectral colors in a rainbow, oil slick, or the reflections from a prism or similar device through which white light is broken up into its individual components.

White (and, for that matter, most colors) is made up of a number of separate frequencies or spectral colors. The eye has the faculty of adding these frequencies in such a way that the colors we actually see are the arithmetical sum of the various individual spectral colors. A good example of this can be observed in stage lighting. A white object is illuminated by one red and one blue spotlight and appears in a purple color. When the red spot light is dimmed, the white object appears more bluish. Similarly, when the blue one is dimmed, the object assumes a more reddish-purple shade. If, in addition to the red and blue, a third—green—light is added, the white object will actually appear white. Though it may be difficult to find occasion to observe this experiment on the stage, actual work with color TV picture tubes will illustrate the fact that by adding red, blue, and green light, white can be obtained.

The apparent addition of colors by the human eye is a well known and widely used phenomenon. It is used in color printing as can be seen by a close examination of the fine pattern which makes up the color pictures in this book. The exact mechanism is still not completely understood, but is apparently based on three or four chemicals in the different receptor cells in the human eye. The combination of all visible and distinguishable colors seems to be performed in the brain, based on the inputs from the receptor cells. While the color mixing has been compared to the mixing action of the mixer stage in a heterodyne receiver, this does not strictly hold true, since the electrical output of the human eye still seems to contain the primary colors which make up the original scene.

How the Eye Sees

To understand the limitations of the eye better, consider Figure 2.2, which shows a simplified schematic cross section of the human eye. The light passes through the color slide and is focused by the lens to fall on the retina, the light-sensitive surface of the inner eye. Between the lens and the retina is the pupil, a circular hole determining the amount of light that can enter the inner eye. During daylight the pupil normally is smaller than at night or when less light is available. The familiar sensation of being blinded by a sudden burst of bright light is due to the fact that the pupil was opened and then took a few seconds to adjust for the brighter light. Naturally, the position of the pupil also affects the position of the lens. When fairly bright light is observed the pupil will be partly closed and the lens will focus only a narrow beam of light into the inner eye. This narrow beam will be concentrated on a relatively small position of the retina; this small area is called the foveal pit. Concentrated around the foveal pit are a large number of rods and cones, the actual photoelectric pick-up devices in the eyes. Each of these cones and rods is connected to fine transparent nerves, which are gathered up into a sort of cable, the optical nerve, which goes to the brain.

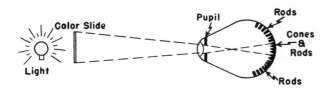

Figure 2.2—CROSS-SECTION OF THE HUMAN EYE

Limitations of the Eye

Although medical science has not yet fully explained the complex mechanism of sight, it is definitely established that the cones furnish color perception while the rods only receive variations in brightness. This agrees with other observations concerning color blindness, night vision, and color perception in general. Since the rods are located all over the retina and the cones are concentrated near the small foveal pit, good color vision is possible only when the light enters in a bright, narrow beam. We know that at low light levels, such as on a moonlit night, the pupil is wide open and while we can clearly distinguish objects, all colors appear indistinct as some silvery shade. The old saying about all cats being gray at night is certainly borne out by the latest biological research.

For a clear understanding of the principles of colorimetry explained in this chapter it is important to remember that color vision depends greatly on the illumination. In a brightly illuminated scene we can distinguish colors better and see many more different shades than in a dimly lit picture. When no light is present, we see black, the total absence of any color.

In addition to brightness variations, various colors themselves react differently on the eye· Some colors are more easily seen than others. A curve showing the relative sensitivity of the eye with regard to the spectral colors is shown in Figure 2.3. Actually, each person has a slightly different color vision and the curve shown here is a result of tests on a large number of people and represents what is called "the standard observer" as defined by the ICI (International Commission on Illumination). From Figure 2.3 we see that the eye is most sensitive to the green shades. If equal amounts of light of the various colors were projected we would have the sensation that green is the brightest of all. This factor is taken into consideration in color television in the composition of a white made up of unequal amounts of other colors, to account for both color vision and picture tube characteristics.

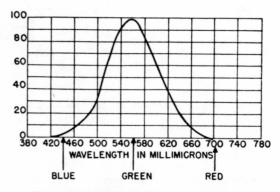

Figure 2.3—RESPONSE CURVE OF THE EYE

A further important aspect of color vision is the ability to distinguish different shades of one particular color. Experiments have shown that the eye has much more difficulty distinguishing similar shades of green than, for example, shades of yellow. This limitation of the eye permits good color reproduction even though not all possible shades of green are reproduced and, as is shown later, allows excellent color fidelity even though some of the true spectral colors are omitted.

There are many more visual limitations and peculiarities that find application in illumination and colorimetric work, but for the purpose of the color television student the factors listed above are the most important.

Color Matching

Anyone who has ever tried to match a new pair of pants to an old jacket will know how difficult it is to match colors exactly. The best way

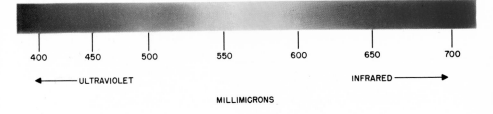

400   450   500   550   600   650   700

◄──── ULTRAVIOLET   INFRARED ────►

MILLIMICRONS

Figure 2.1—SPECTRUM CHART

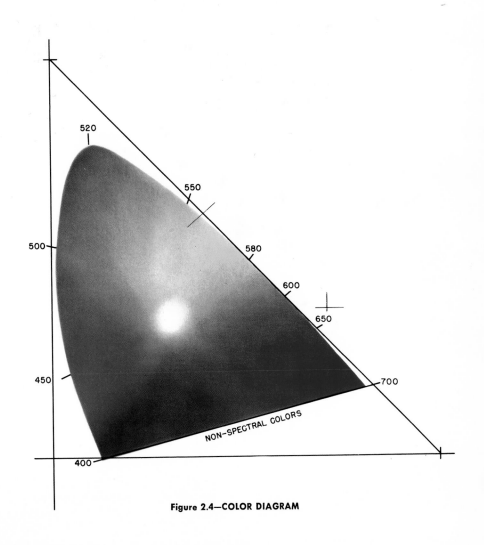

Figure 2.4—COLOR DIAGRAM

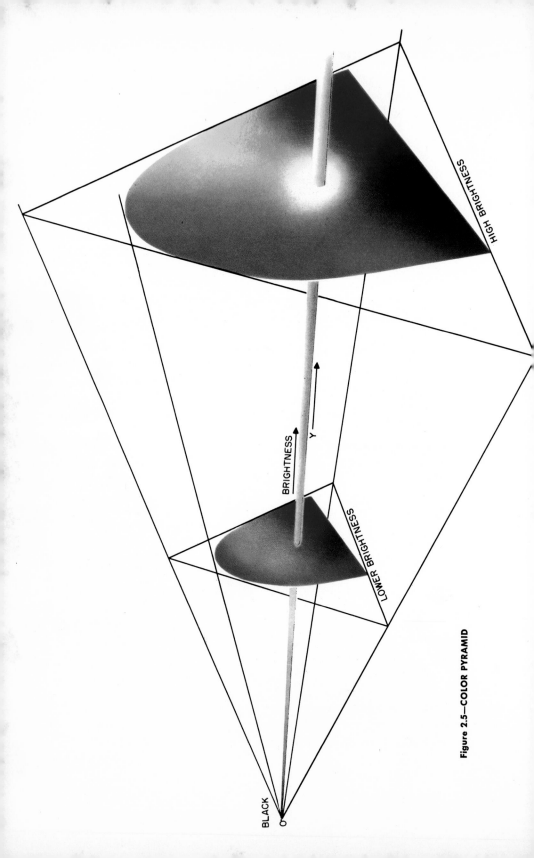

Figure 2.5—COLOR PYRAMID

to do this is to hold the two materials close together and even then it is often found that a good color match under artificial light appears a poor match in sunlight. This problem of color matching is one of the basic reasons for the science of colorimetry, and much time and effort has been devoted to it. In industry the color of paint, dyestuff, beverages, and so forth, is of great importance and many instruments have been developed for accurate measurements. Doctors determine the red corpuscle content of the patient's blood by color-matching with calibrated samples; similar comparison methods are used to detect diabetes, kidney trouble, and other conditions. It is obviously necessary that certain standards be set for colors and that a system of defining colors can be established to assure that, regardless of the visual limitations of the individual observers, colors are really identical.

A Color Coordinate System

Space is measured by giving height, length, and depth; radio signals are defined by frequency, amplitude, and waveshape; and colors are similarly defined by three quantities. Although there are several different systems of color measurement, a parallel to the metric, inch, and pound systems, we shall only consider the one used in color television. This system defines a color by these three aspects:

Brightness—the amount of light, regardless of coloring.

Hue—the predominant color such as red, green, yellow.

Saturation—the degree of purity of the color. Since a single-freqency color rarely occurs alone, saturation determines the amount of the other colors or the amount of pastel shading.

There are several mathematical transformations that connect the various color-measuring systems and these formulas are used to convert a simple, uniform brightness color presentation into a more useful color diagram such as the one shown in Figure 2.4. The basic formula used involves a plot of all spectral colors when it is assumed that the sum of the red, green, and blue color coefficient is equal to one. From this principle the color triangle was evolved and it represents all theoretically possible colors and their mixtures at one given brightness level. The color print (Figure 2.4) uses only those colors that are reproducible by a good color printing process and substitutes them for the natural colors. Color television is capable of reproducing many more colors than this printing process, and nature presents a still more varied color gamut. This effect is similar to the sounds occurring in nature, audible through a small radio and through a high fidelity music system.

Specifying Colors

The color diagram (Figure 2.4) shows some of the principles of color adding and mixing. Spectral colors are located along the horseshoe-shaped rim of the diagram and these are theoretically pure, or have 100 per cent saturation. White lies in the center and is always a mixture of at least three other colors; its saturation is zero. When looking at Figure 2.4 we can readily determine the *hue* of any color. It is that spectral color that lies at the radius from white to the outer rim. This suggests the use of an angular measure in defining hue. If a certain red, 650 in Figure 2.4, were considered as reference line, the hue of some color could be given as 45 degrees from the reference. In color television this angular measure for hue is quite important. In order to define a color clearly, both its hue and its saturation must be given. When hue is specified by an angle, the degree of saturation can be specified by stating the distance from the white center. Referring again to Figure 2.4, a color can be defined by its hue, and its distance from the center, which is the white point in the color diagram. From this we can see that any color in the diagram (Figure 2.4) can be specified by the hue-angle and the saturation-distance. These concepts may seem somewhat artificial in this connection, but they form the basis for the electrical signal specifications used in color television.

In addition to giving hue and saturation, colors can also be specified by rectangular coordinates, referring to the horizontal and vertical distance of a point from the origin of the dotted-line axes. This system of determining colors is used in colorimetry and forms the mathematical basis for the hues and saturation system used in color TV.

So far we have assumed that all colors have the same brightness, but this is not true in nature. Illumination, shadows, different reflection from different materials all cause colors to appear in identical hues and saturation, but with different brightness. Observe a yellow material under bright light and when a shadow falls across it. The effect of the shadow tends to make the yellow more of a light brown. Similarly, a white material appears gray when less light is supplied. In order to reproduce colors accurately in color TV the color brightness must be reproduced. This color brightness is called *luminance* to distinguish it from the black and white brightness. Luminance is always understood to be associated with a certain color, and it is possible to specify a color by its luminance, hue, and saturation.

One system of color coordinates widely used in the food industry assigns a set of numbers to a color that completely determines it in all respects. These are called the *tristimulus* values, because all three aspects, brightness, hue, and saturation, are taken into account. Since the tristimulus system is not used in the final color TV signal standards, it is sufficient to know about its existence without going into mathematical details.

Color Pyramid

If the color diagram (Figure 2.4) contains all colors of equal brightness, the sum total of all colors of all brightnesses must appear as a solid pyramid, somewhat like the one shown in Figure 2.5. Although only two color diagrams of different brightness are shown, the actual color pyramid consists of an infinite number of such planes stacked together to form all possible colors at all possible brightness levels. As the brightness increases, more color detail is visible and the color diagrams become larger. At low levels of brightness, the colors become less distinct and the color diagrams shrink. When brightness is zero no colors can be seen; therefore the zenith of the pyramid represents black. Theoretically, at infinite brightness the colors are indistinguishable and only white appears to the eye, but ordinarily we would have to close our eyes long before this tremendous brightness and for all practical purposes bright sunlight or the Klieg lights used in TV studios supply maximum brightness. The color pyramid is not an actual body found in nature, but is merely a geometric representation of the characteristics of the eye and the colors it sees.

The effect of brightness variation on the white center of the color diagram is shown by the narrow cone going from black at the zenith through all shades of gray to the white of the brightest color plane. This cone is often called the Y or brightness axis, because distances along it are used to indicate brightness levels. In monochrome TV only variations in brightness or various shades of gray are reproduced. The TV signal on monochrome transmissions therefore corresponds to the Y axis. For color TV the whole pyramid made up of the three primary colors is reproduced. Therefore the color TV signal must specify any given point within this pyramid. Here is how this is done.

One signal, the monochrome brightness signal, corresponds to the Y axis and thereby locates the color plane in which the particular color is located. A second signal indicates the angle from a reference line and a third signal corresponds to the distance from white at which the color is located. In Figure 2.5 color A could be specified as 10 divisions along the brightness axis, 45 degrees from the reference line, and 3 divisions from the white center. Another way of stating this would be to say 10 volts of brightness signal, 45 degrees phase shift from the reference voltage, and 3 volts of saturation signal. Thus by assigning electrical quantities to hue, saturation, and brightness, a color is determined, transmitted, and reproduced.

Color Coordinate Transformation

The original statement that any color can be reproduced by a suitable mixture of three primary colors is still true. Although we have chosen brightness, hue, and saturation as a coordinate system, the color space model (Figure 2.5) can be considered as having any other set of three-dimensional coordinates, including three primary colors. The brightness, hue, and saturation coordinates have been derived mathematically from a set of primary color coordinates and, in the matrix sections of trans-mitter and receiver, the transformation is carried out electrically. This is necessary because both the camera and the picture tube must deal with what the eye sees and relate these actual colors into electrical signals. The various electrical transformations are necessary for the TV transmission and are explained in more detail in the next chapter. When the color television signal was formulated, the coordinate system shown in the color pyramid was used because in this system the Y or brightness coordinate is completely independent of color and can be utilized for black and white television.

One of the features that allows color TV transmissions to be re-produced in monochrome receivers without any adjustment or converter is the use of the Y or brightness signal. In other words, a color receiver can receive, detect, and amplify all three signals necessary for a color picture while the monochrome TV set detects only the brightness signal and reproduces the identical picture in black and white. Conversely when a monochrome TV transmission is received, both the color and monochrome TV set operate only on the Y signal and both produce black and white pictures. The color camera at the transmitter produces three signals, the red, green, and blue voltages. This corresponds to the red and green and blue primary coordinates of the color pyramid. To change these signals into signals corresponding to the Y, hue, and satura-tion coordinates, the voltages are changed as shown below. Before going into the arithmetic of these equations, remember that they concern AC voltages of varying phase and amplitude. Addition, subtraction, and multiplication are carried out by amplifiers, adding networks, and phase shifting circuits.

Brightness-signal $E_Y = 0.3E_{red} \quad + 0.59E_{green} \quad + 0.11E_{blue}$

Hue $E_Q = 0.21E_{red} \quad - 0.52E_{green} \quad + 0.31E_{blue}$

Saturation $E_I = 0.6E_{red} \quad - 0.28E_{green} \quad - 0.32E_{blue}$

These relationships effectively transform the red, green, and blue signals from the color TV camera into the Y, I, and Q signals used for color transmission. At the receiver, the same relationship is used to produce again red, green, and blue signals for the color picture tube.

3

COLOR TV SIGNALS

From the preceding chapter we know that three different signals are necessary for color transmission. One of these signals is the brightness signal, which is essentially the same as for monochrome TV reception. This brightness signal determines the various shades of gray in black and white TV and its frequencies range from about 100 Hz to almost 4 mHz. The higher frequencies correspond to fine lines, edges, or other details in the picture. In order to reproduce all these fine details the brightness signal bandwidth in a good monochrome receiver should extend to 4 mc. In color TV the same requirement applies to the brightness signal since changes in brightness are more perceptible than changes in colors, especially in small areas. The eye is also fairly sensitive to rapid changes in the amount of white contained in a color, or the degree of saturation, while it is least sensitive when it comes to difference in hue. These limitations of the eye are utilized in determining the bandwidth of the various color signals. After extensive tests of the color vision of many individuals the bandwidth of the three color signals necessary for good pictures was established. The final values are given below:

Brightness	Y signal	4 mHz
Saturation	I signal	1.3 mHz
Hue	Q signal	0.4 mHz

Although these are the bandwidths of the three transmitted signals, the receiver characteristics will determine the actually reproduced bandwidth and the possible degree of color fidelity.

In color TV the Y or brightness signal is transmitted in the same manner as the monochrome video signal. The major problem is to transmit the I and Q signals, together with the Y signal in the same 6 mc-wide channel. Before going into the details of how this is accomplished, let us consider the possibilites of modulating a single carrier with two different independent signals. One method of modulation is amplitude modulation, which is used for the monochrome video and radio broadcast signals. In this method the amplitude of the RF carrier is varied in accordance with the audio or video signal. Another modulation process with which the reader will be familiar is frequency modulation, where the frequency of the RF carrier is varied in accordance with the audio signal. The sound accompanying TV signals is frequency modulated, for example. A third method of modulation is called phase modulation. As the name implies, the phase of the RF carrier is changed in accordance with the video or audio signal. In order to use phase modulation it is necessary to send out a reference phase signal together with the modulated RF signal. Phase modulation and reference phase signals are used to transmit the hue information in color television and it is necessary clearly to understand the various phase relationships. Before delving into this subject the reader may want to refer to the following paragraphs, which present a brief review of the principles of vectors and vector presentation.

Principles of Vectors

A vector not only describes a quantity, but also deals with some action. When a vector is used to describe the pull on a rope, the force in pounds as well as the direction of the pull is given. Looking at Figure 3.1 we see that the angle of the incline on which the weight rests is as important as the size of the weight itself when it comes to balancing it with a counterweight. The force of the weight is therefore given by the vector W and the angle A.

Figure 3.1—VECTOR PRINCIPLE

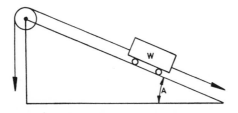

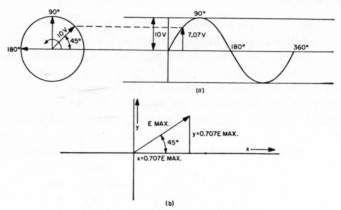

Figure 3.2—ELECTRICAL VECTOR

An application of vectors in electricity is illustrated in Figure 3.2 where phase is the angular measure, and voltage amplitude corresponds to the magnitude of the vector. The complication in the electrical system is that the phase does not stand still, but, in sine wave signals, represents a rotating vector that moves through a complete circle in the period it takes the signal to pass through one cycle. When the vector at the left of Figure 3.2 rotates through a sine wave voltage cycle, the radial vector can be defined either by giving its angular position and its length or by giving its vertical and horizontal components. Thus vector R at the 45 degree point can be described as "10 volts, at angle 45 degrees" or it can be described as "0.707 horizontal plus 0.707 vertical." The first notation would be $E_{max} \angle 45$ degree and is in polar coordinates. The second notation can be written as

$$\frac{x\sqrt{2}}{2} + \frac{y\sqrt{2}}{2} = (x + y) \frac{\sqrt{2}}{2} = 0.707 (x + y)$$

The latter notation is in Carthesian coordinates, often called orthagonal coordinates or simply x, y coordinates. In order to transform from one type of coordinate system to the other the following equations are helpful:

$$E_{max} \angle a = E_{max} \sin a + E_{max} \cos a = x,y$$
$$E_{max} = \sqrt{x^2 + y^2}$$
$$a = \arctan \frac{x}{y}$$

where a = angle of vector with x axis
E_{max} = maximum magnitude of vector
x = horizontal coordinate
y = vertical coordinate.

20

The relationship of Figure 3.2b is quite important and is based on the geometry of a right triangle, the sine and cosine functions. It will not be necessary for the technician to perform any algebra involving vectors in color TV service work, but the mathematical principles must be understood in order to follow the explanation and to comprehend the various color TV signals and matrixing operations. These operations involve adding and subtracting vectors by means of electrical circuits; their mathematical equivalent of vector addition is described below.

Because a vector contains both magnitude and direction, two vectors with different directions cannot be simply added numerically. Figure 3.3

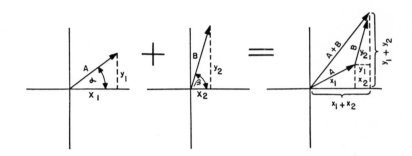

Figure 3.3—VECTOR ADDITION

shows how two vectors, A at angle a and B at angle β, are added graphically to form one vector A *plus* B at a completely new angle. The right-hand portion of Figure 3.3 also illustrates the fact that while neither the magnitudes nor the angles can be simply added, the x and y coordinates of each vector can be added arithmetically. To express the addition algebraically the two vectors are written as follows:

$$A \angle a + B \angle \beta = (x_1 + x_2) + (y_1 + y_2)$$

Subtraction of two quantities is nothing more than an addition where one item has a negative polarity. This is the same as the operation of negative grid bias voltage in a tube circuit which is overcome by a positive polarity signal. Arithmetically the subtraction of vectors is done just as addition, by subtracting the two x-coordinates and the two y-coordinates from each other.

$$A \angle a - B \angle \beta = (x_1 - x_2) + (y_1 - y_2)$$

The graphical presentation of vector subtraction requires that the direction of the negative vector be reversed. This is shown in Figure 3.4 which illustrates in *a* the addition and in *b* the subtraction of vector A

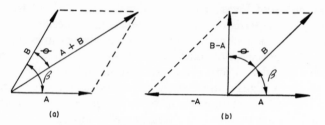

Figure 3.4—VECTOR ADDITION AND SUBTRACTION

from B. In this illustration we have considered vector A as being on the reference axis and therefore having no y component. Angle a in this case equals zero. Another change from Figure 3.3 lies in the fact that the two vectors are not of equal size. This illustrates further the fact that magnitude or angle cannot be added or subtracted directly in vector algebra. The practical application of vector analysis in color TV lies in the fact that signals of equal frequency but different phase angle must be added and subtracted to produce a certain color signal.

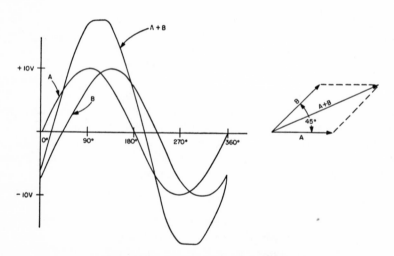

Figure 3.5—ADDING TWO SINE WAVES

Figure 3.5 illustrates how two sine waves, A and B, which have a phase difference of 45 degrees, are added to produce a third wave form. Note that at each point of the third wave the total amplitude is the arithmetical sum of the amplitudes of the two other waves. Where a sine wave goes below the zero axis, it has a negative value and its amplitude is subtracted from any amplitude above the zero axis. A vector presentation of the wave form is shown at the right of Figure 3.5 and it should be remembered that this illustrates the relationship of the three signals at each instant and that all three vectors are rotating together at the frequency of the signal. To simplify the explanation we have assumed that voltages A and B are of constant 10 volts amplitude and have a constant phase difference of 45 degrees. An actual color signal will consist of two such voltage vectors in which both the amplitude of each and the phase angle between them will change constantly. The resultant waveform will be quite irregular. If one of the two voltages is of constant amplitude and phase, while the other one varies the

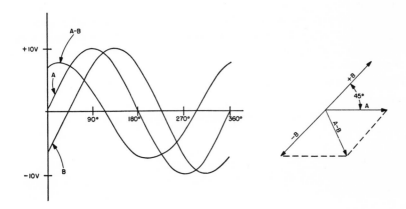

Figure 3.6—SUBTRACTING TWO SINE WAVES

result will be the demodulated video information contained in the varying signal. As an example of the subtraction of two sine wave voltages A and B, the same amplitudes and phase angles are used for the addition; the result is shown graphically in Figure 3.6. Just as in the addition, the amplitude at each point of the resultant curve is the total of the subtraction of the two original waves A and B.

In both Figures 3.5 and 3.6 we have assumed that the two sine waves A and B are equal in amplitude and have a constant phase difference of 45 degrees, and the results we get are far from the apparent arithmetic sum or difference. In the case of the addition the resulting waveform is no longer a true sine wave—harmonics have been introduced.

It is possible to modulate an RF signal by shifting its phase with respect to a certain reference signal of the same frequency. Then, if the detector is sensitive to the phase difference between the two signals, the original audio or video signal can be recovered. This is the principle of the method of transmitting the color information in color television and a thorough understanding of phase modulation and phase detection is therefore required of the serviceman.

Color Sub-carrier

As mentioned before, the Y or brightness signal in color TV is practically the same as for monochrome. The I and Q signals are both transmitted on a 3.579 mHz sub-carrier. In the following section the method of sending this sub-carrier along with the regular TV signal will be described in detail, but for the moment only the color sub-carrier itself is considered. It would be rather simple to amplitude modulate the sub-carrier with either the I or Q signal, but since both of these signals must be transmitted, phase as well as amplitude modulation is employed. The method of modulating or encoding the color sub-carrier can be considered as using amplitude modulation for the I signal and phase modulation for the Q signal. The main feature of the system is the fact that a 90 degree phase angle exists between the I and the Q carrier. In addition to this phase difference, a 57 degree phase angle exists between the I and the reference or zero phase signal. The relation of these voltages is shown in the vector presentation of Figure 3.7.

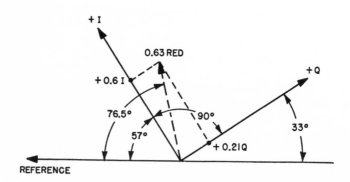

Figure 3.7—*I* AND Q, AND A COLOR VECTOR

To give an example of how these two signals, the I and Q, are used to represent a certain color, we have chosen 0.63 volts of red. This color will be produced when the I signal amplitude is 0.60 volts and the Q signal is 0.21 volts, both of positive polarity. With this combination of I and Q signals the matrixing circuit at the receiver will produce 0.63 volts of red. When other values of I and Q with different polarity appear, they signify different colors. Figure 3.8 shows the combination of I and Q signals required to produce 0.45 volts of blue.

It should be pointed out that the magnitude of the vectors here are taken as relative rather than absolute values. Thus it may be that at some level in the matrixing circuit the unit vector of I and Q is 10 volts, or 30 volts or some other value. In each instance the values of I and Q required to produce the 0.63 red shown in Figure 3.7 would be 10 or 30 times the 0.60 I and 0.21 Q signal shown in Figure 3.7.

In Figures 3.7 and 3.8 only the primary values of red and blue are indicated, but between them, in angular displacement, lie all the shades of purple, violet, and magenta which the color diagram provides. The angular values of the I and Q signals themselves will correspond to a certain shade of orange and magenta respectively.

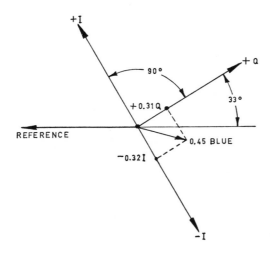

Figure 3.8—COMBINATION OF I AND Q TO GIVE BLUE

The principles of colorimetry which allow us to separate the Y or brightness signal were discussed in Chapter 2. One of the conclusions drawn then was that by transmitting the Y signal and two of the color difference signals, the total color information could be sent out. In some receivers the matrixing or decoding sections operate on those two color difference signals rather than using the I and Q signals and for that reason Figure 3.9 shows the vectorial relationship of these signals. The R-Y, or red difference signal, is displaced 90 degrees from the reference. I and Q signals are located as shown in the diagram.

Most of today's color TV models do not use the I & Q reference vectors in their demodulators. They use either the red and blue difference signals or else two arbitrary vectors, chosen for convenience for circuitry, and called the X and Z vectors as shown in Figure 3.9. Regardless of the type of receiver used, however, the transmitted signal is based on the geometry and the bandwidth requirements of the I and Q vectors.

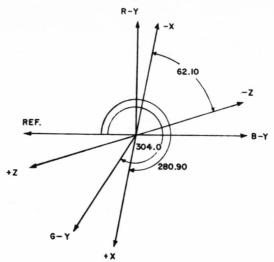

Figure 3.9—COLOR DIFFERENCE VECTORS

The bandwidth of the I and Q color sub-carrier was determined by a number of related factors. If I is taken to represent saturation and Q the hue, the range of variation in saturation, the bandwidth required for I, is found to be greater than that for changes in hue. This is based on the sensitivity of the eye to variations in hue and in saturation. Considering only colors, I represents orange-yellow and Q magenta, and the human eye is more sensitive to detail in orange-colored objects than in magenta-colored ones. Based on these and other, more involved, considerations the bandwidth of the I signal is set at 1.2 mHz and the Q signal at 0.4 mHz at the 3 db point. This is illustrated in Figure 3.10, which shows that the Q signal has full double side band modulation while the I signal has one of its side bands suppressed in part. It is important to keep these bandwidths in mind when considering the bandpass responses of the various tuned networks in the TV receiver.

Modern color television receivers usually do not utilize the full bandwidth provided by the 1.2 mHz I signal. Tests with untrained observers have shown that the additional bandwidth obtained in the orange region is not sufficiently noticeable to most viewers to warrant

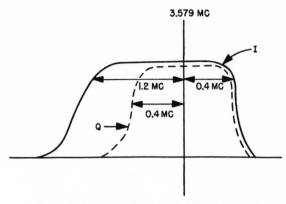

Figure 3.10—BANDWIDTH OF I AND Q SIGNALS

the extra cost required for the special bandpass circuits. For this reason practically all modern receivers use a ± 0.5 mHz bandwidth at the −3 db point and rely on the R-Y, B-Y or on the X & Z vectors for demodulation.

Frequency Interlacing

One of the features of color television is that each channel occupies the same mHz bandwidth as the monochrome transmission. Yet the color signal contains brightness information with 4 mHz bandwidth, color information with a total of over 1.7 mHz bandwidth, and the sound carrier located 4.5 mHz from the video carrier and having at least 0.1 mHz width.

The utilization of the 6 mHz channel for the color sub-carrier is possible because of a peculiarity of the video signal. Close examination with a spectrum analyzer reveals the fact that the video signal from 60 Hz to 4 mHz does not consist of a continuous band of energy, but rather is made up of bursts of energy, spaced 30 cycles apart. This is due to the 30 cycle scanning rate, the lowest repetitive frequency in the video signal. Actually the bursts of energy also occur at 60 Hz and 15,750 Hz, other scanning rates which are multiples of the 30 Hz frame rate. Figure 3.11 shows in exaggerated form how a typical video signal consists of individual bursts and how the space between bursts is utilized for the color signal. Since the spacing and repetition rate of the individual burst of energy depends on the scanning of the picture, it is possible to insert another signal between these spaces by making certain that the second burst occurs at the same rate, displaced by 15 cycles.

This system of transmitting the color information along with the Y signal is called frequency interlacing and is one of the important principles new and peculiar to color TV. To make sure that each of

27

the energy strips of color information falls between two strips of Y signal, the color sub-carrier frequency is chosen to be an odd harmonic of half the line scanning frequency. The exact figures are 3.579545 mHz, which is the 455th harmonic of one half the line frequency. Up to now we have given the color sub-carrier as 3.579 mHz because the exact value is not important to the service technician. For simple reference the 3.58 mHz notation is often used. Only at the transmitter must the various frequencies be related exactly and at the receiver the synchronization and decoding circuits automatically reproduce the proper frequency relationship.

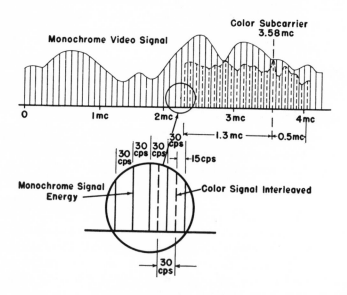

Figure 3.11—FREQUENCY INTERLACING

The signal shown in Figure 3.11 can be received by any monochrome receiver and the effect of the 3.579 mHz color sub-carrier will not be noticed on the screen. There are no provisions for detecting or removing the color sub-carrier and since its strips of energy fall in the spaces ordinarily left blank, no information will appear on the screen due to the sub-carrier. In some receivers it is possible that a slight 3.579 mHz beat will appear due to some non-linear circuit in the set or due to overemphasis of the higher video frequencies.

In color receivers the decoding section separates the color sub-carrier from the video signal and because synchronous detection is used, only the color information is extracted. The separation of the color sub-carrier from the other information and the detection or demodulation of this carrier must be made by comparison with a reference signal of 3.579 mHz. This reference signal synchronizes the color sub-carrier with the sub-carrier at the transmitter in the same basic manner as the horizontal synchronizing pulses lock in the horizontal scanning frequency at the receiver. For this reason we refer to the 3.579 mHz reference signal as the color synchronizing signal.

Color Synchronizing Burst

Just as the vertical and horizontal synchronizing pulses must be sent out along with the video signal, so must the color synchronizing signal be sent out together with the frequency interlaced color sub-carrier. The method of doing this makes use of the trailing edge of the horizontal blanking ledge as shown in the somewhat exaggerated view of Figure 3.12. During the period of this blanking ledge, while the electron

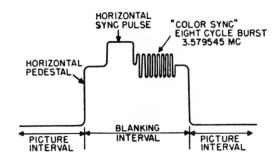

Figure 3.12—COLOR SYNCH BURST

beam is blanked out and returns to the left side of the screen, a short burst of eight cycles of 3.579 mHz sine wave appears. This burst, often called simply the color burst, is a short portion of the continuous 3.579 mHz reference sine wave voltage which is used at the transmitter to encode the color signal on the sub-carrier. At the receiver the color burst serves to lock in a local 3.579 mHz sine wave oscillator which generates the equivalent of the transmitter reference voltage. Later chapters will provide a more detailed explanation of just how this reference local color oscillator works, how it is synchronized and how its output serves to demodulate both the X and Z signals.

The complete color TV signal consists of three separate video signals, and three separate synchronizing signals. A frequency relationship exists between the three synchronizing signals since the horizontal synchronizing frequency is a harmonic of the vertical and the color synchronizing signal an odd harmonic of half the horizontal frequency. Black and white information is transmitted on a regular amplitude-modulated RF carrier with 4 mHz bandwidth while the two color signals are phase and amplitude modulated on the 3.579 mHz color subcarrier. In addition to these video and synchronizing signals the frequency modulated sound carrier is transmitted and located 4.5 mHz above the Y or main video carrier. Figure 3.13 shows the frequency

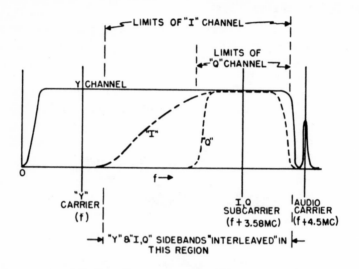

Figure 3.13—FREQUENCY OF ALL SIGNALS IN 6 mHz CHANNEL

distribution, bandwidth, and separation of the video carrier, color subcarrier, and the FM sound carrier. Note that the separation between the sound carrier and the partially suppressed sideband of the color sub-carrier is relatively small and that a possibility for cross modulation and interference exists. This problem requires that the stability of the local RF oscillator in the TV tuner be fairly good and that sharp bandpass responses are employed in the receiver IF section. How these requirements are met in a typical color TV receiver is described in a later chapter.

Summing up the salient features of the color TV signal as compared to monochrome transmissions we find that the following new principles have been introduced:

1. Phase and amplitude modulation. Use of both types makes it possible to modulate the color sub-carrier with the hue as well as the saturation video signal.

2. Color sub-carrier. In addition to the regular RF carrier a sub-carrier at 3.579 mHz is used.

3. Unequal side bands. As becomes evident from study of Figure 3.13, it is necessary to limit the upper side band of the sub-carrier to avoid interference with the sound carrier.

4. Frequency interleaving. This principle allows insertion of the color sub-carrier into the unused spaces between strips of energy of the main video signal. The scanning frequency determines the strip repetition rate and by using half of an odd multiple of the scanning frequency, the color sub-carrier strips are placed between the brightness signal strips.

5. Color synchronizing bursts. To transmit a 3.579 mHz sine wave reference signal, a short burst of this signal is placed on the blanking ledge of each horizontal synchronizing pulse. This burst is converted into a continuous sine wave at the receiver and serves to lock-in the demodulating circuits.

A COLOR
TELEVISION SYSTEM

The Camera

Just as in monochrome television, the color TV camera originates the color picture information. Color TV cameras are in effect three separate cameras, each receiving one of the three primary colors, red, green and blue. In actual studio equipment four cameras are often used with three providing only the color information while the fourth acts as high resolution black and white camera. Because of various linearity and level problems, color TV cameras are not exactly the same as their monochrome predecessors. Special camera tubes as well as special filters and compensating circuitry are used. For a basic understanding of color television, however, we can limit our considerations to the three cameras shown in Figure 4.1.

A single scene is viewed by all three camera tubes in Figure 4.1 but the different colors, that reach each of the tubes, depend on the filter placed in front of each camera tube. The red filter passes only the red components, the green only the green components, and the blue only the blue portions of the picture to the respective camera tubes. Following each camera tube is a series of amplifiers just as in monochrome TV, but the gains of the three amplifier chains are adjusted to compensate for any shortcomings of filters and tubes and to produce properly balanced red, green and blue signals. One of the important aspects of the color camera is the fact that all three camera tubes have the same scanning signal applied so that the three electron beams scan the scene in absolute unison. For this reason the block diagram of

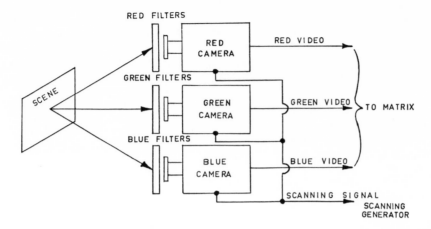

Figure 4.1—BLOCK DIAGRAM OF CAMERA SYSTEM

Figure 4.1 shows a single scanning signal applied to all three cameras. The three separate video signals correspond to the red, green, and blue content of the televised scene.

From the camera the signals usually go through various amplifiers, monitors, switching panels, and the like, until at the transmitter they are fed into the encoder and matrixing unit.

The Encoder and Synch System

Figure 4.2 illustrates how the three color signals reach the encoder and how the color sub-carrier is applied as well as the reference burst derived from it. Three filters control the passband of the Y, I, and Q signals at the encoder output. As stated in the previous chapter the bandwidth of the brightness or Y signal is up to 4 mHz. The I signal has a bandwidth of 0.4 mHz at one side and 1.3 mHz at the other side band and therefore the passband for the I signal is from 2 to 4.2 mHz. The Q signal, when modulated on the color sub-carrier has a passband from approximately 3 to 4.2 mHz.

Note that in the block diagram of Figure 4.2 there are now three video and one synchronizing signal going to the RF modulator. The synchronizing signal carries the vertical and horizontal pulses as well as the color synchronizing burst. Actually the synchronizing signal occurs in the blank spaces of the Y or brightness signal and is simply added to it.

The color sub-carrier and its sidebands cannot be applied directly to the modulator. Since the final color signal will consist of a wide range of frequencies, and since the phase relationships of the individual signal components are so important, care must be taken that this phase relationship is not disturbed in transmission or reception. In the receiver the bandpass responses of the RF and IF section cause different phase delay at different frequencies and further vary the phase relationship between the color sub-carrier and the brightness video signal. To overcome these limitations in part, the transmitter contains a section called "phase delay equalizer" which consists of various filters, phase shifting networks and bandpass amplifiers. This section of the transmitter compensates for phase delay errors due to antenna and receiver characteristics and equalizes the video signal before it reaches the RF modulator.

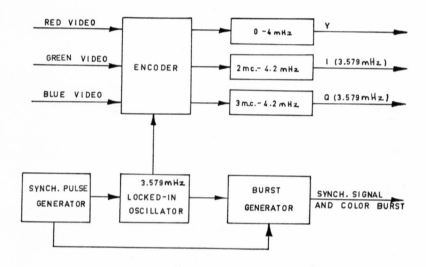

Figure 4.2—ENCODER BLOCK DIAGRAM

In actual color TV studios there is one more feature that is not found in monochrome TV stations—the need for the gamma correction amplifiers. Gamma is the name given to the relationship in brightness which various shades of gray will have on a given picture tube. Ordinarily, equal increments of signal voltage at the kinescope grid do not produce equal increments of brightness near the maximum brightness point. This is due to the non-linear grid plate characteristic of the picture tube. In most TV receivers there is some compensation for this in the design of the preceding video amplifier which displays a similar characteristic, but this tends to work in reverse since the video signal is inverted at the grid preceding the kinescope. In any event, gamma errors in monochrome receivers are not a serious problem. They are, however, of concern in color TV. Just as in the case of the phase delay equalizer, the errors due to non-linearity in the receiver are compen-

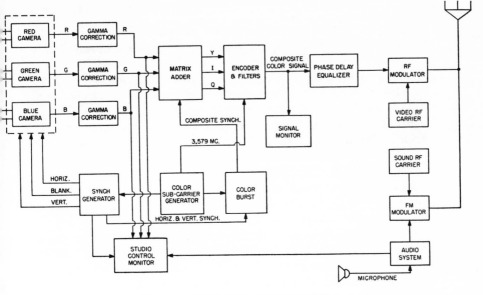

Figure 4.3—TRANSMITTER BLOCK DIAGRAM

sated for at the transmitter. This system permits the station to compensate also for non-linearity in its system, especially in the camera tubes.

As shown in Figure 4.3, the three separate gamma amplifiers follow the cameras and make the necessary corrections on each signal before the encoding begins. The technician need not be familiar with the details of gamma amplifiers but it is well to know that they are essentially amplifiers with a non-linear element as plate or grid load. By choosing the right type of load, usually a diode, the amplitude response of the circuit can be adjusted so that the ordinary grid-plate characteristic is inverted and that larger grid signals get more amplification than smaller ones.

The Complete Transmitter

A block diagram showing the major portions of the color TV transmitter is shown in Figure 4.3. The portions which differ from the monochrome TV transmitter are the camera section, gamma correction amplifier, color synch signal generator, encoder, and phase delay equalizer. RF modulation, audio, and the FM sound carrier portions are essentially the same as for monochrome transmissions. The various line, distribution, and monitoring devices differ from the monochrome version mostly in that the specifications for each of them are more

35

severe, especially as concerns bandwidth and phase delay. Another feature is that for all video circuits three lines are required instead of only one.

Analyzing the simplified color TV transmitter shown in Figure 4.3 we note that the color camera is controlled by the horizontal and vertical scanning signals from the synch generator as well as a blanking signal that eliminates any picture pick-up during the retrace period. Each of the three individual camera tubes produces a separate video signal corresponding to the red, green, and blue light components of the televised scene. These video signals pass through the gamma correction amplifiers, which compensate for non-linearities in the transmitter as well as receiver circuitry. Gamma correction is necessary to reproduce a full range of grays, often called the gray scale. From the gamma amplifiers the three video signals are piped to the studio control and monitoring facilities. Switching from one camera to another, manual color compensation, various directions, adjustments, and other required control operations are performed on the three signals at the studio control console before matrixing begins.

At the matrix and adder section the R, G, B, signals are changed into Y, I, and Q values, still as separate signals with full 4 mHz bandwidth each. The relationships of the R,G,B and Y,I,Q values in arithmetic form were given in Chapter 3; electronic circuits perform the arithmetic accurately and instantaneously. The principles of adding or subtracting voltages of either polarity will be familiar to the reader from the understanding of how grid bias is overcome or added to a signal. A detailed discussion of matrixing circuits will be found in a later chapter on receiver matrix networks.

At the matrix section the three signals are also supplied with the horizontal, vertical, and color synch information. The latter consists, as shown in Figure 3.2, of a short burst of 3.579 mHz sine wave, superimposed on the trailing blanking ledge following each horizontal synch pulse. To obtain this burst, the color sub-carrier generator is locked in with the horizontal and vertical synch generator and feeds a sine wave signal to the color burst section. This section has a gating network, actuated by the horizontal pulse, which chops the 3.579 mHz sine wave into bursts. A simple adding circuit then superimposes the burst on the correct blanking ledge. The encoder and filter section is actually the sub-carrier modulator. The final encoded signal is produced in this part of the transmitter. Actually the Y signal passes through the encoder unchanged, except for amplification and bandpass correction, while the I and Q video signals modulate the 3.579 mHz color sub-carrier. The type of modulator circuit used is somewhat complex, but both amplitude and phase modulation take place. Usually a balanced modulator system is used to obtain the 90 degree phase difference between the I and Q signals. Circuit details of these networks are not important for the serviceman but an understanding of the principles of operation will be gained from the later chapter dealing with decoders and synchronous detectors as used in color TV receivers.

A final signal monitor is used to make certain the signal is properly encoded and will produce correct color pictures. As mentioned before, a phase delay equalizing network is required before the composite video signal reaches the RF section. This phase delay correction network consists generally of bandpass amplifiers with a frequency response curve designed to compensate for the variations in receiver response curves. Also included in this section are time delay networks about which more will be said in the chapter on decoders in TV receivers.

The RF modulator, video RF carrier, and the entire sound section are all essentially the same as for a monochrome TV station. Since the sound RF carrier is always exactly 4.5 mHz higher in frequency than the video carrier, some interlocking frequency control system is usually used to insure good sound reception at the receiver. In color TV this relationship must be maintained even more accurately to avoid beats between the color sub-carrier and the sound carrier.

In the block diagram of Figure 4.3 we have omitted various studio instruments used for tests, switching panels, such additional signal sources as flying spot scanners and film pick-up equipment. Commercial announcements are often on film or in the form of color slides and this is where flying spot scanners and film scanners are used. The signals derived from any of these sources are the same R,G,B format as those described above and processed in the same manner as shown in Figure 4.3. Usually, however, separate gamma amplifiers are used with each of these various signal sources since the color sensitivity and pick-up characteristic varies between them.

The Color Receiver

Before studying the detailed operation of the color TV receiver, the overall functions of the major sections should be understood. The block diagram of Figure 4.4 shows these various sections and the relationship they have to each other. Starting at the upper left, the antenna and RF tuner is essentially the same as for monochrome TV sets. The IF amplifier and detector sections are also functionally the same as for monochrome receivers. In the IF section, the video and sound IF carriers and all the color sub-carrier information is amplified from the

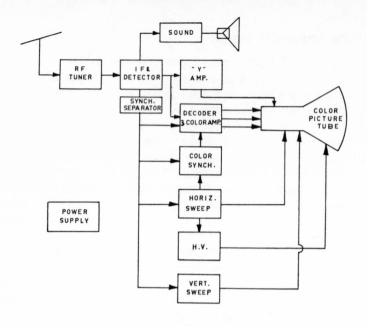

Figure 4.4—RECEIVER BLOCK DIAGRAM

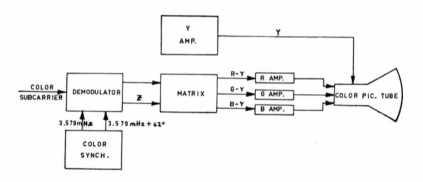

Figure 4.5—DECODER BLOCK DIAGRAM

weak signal at the antenna to one strong enough to drive the video amplifier stages. The sound carrier is at 4.5 mHz and the audio signal must be detected by some type of FM detector.

The composite video signal present at the detector output contains the vertical and horizontal synch pulses, the color synch burst, the Y or brightness signal and also contains the color sub-carrier with the I and Q signals modulated on it. To separate all these different signals and direct them to their respective sections requires many different networks, amplifiers, clippers, and so forth. For purposes of simplicity they are all combined in the one section called "Synch Separator."

38

A bandpass amplifier separates the color sub-carrier and its side bands from the other signals and supplies this 3.579 mHz signal to the decoder section. At the same time the color reference signal is generated in the color synch section and locked in with the color synch burst. To remove the burst from its location on the horizontal blanking ledge, a keying signal, which comes from the horizontal sweep section, is required. Both the horizontal and vertical sections are synchronized by their respective synchronizing pulses in much the same manner as in monochrome receivers. The high voltage supply is usually of the flyback type used in monochrome receivers, but voltages, currents, and components are quite different in magnitude and required precision.

A major difference between the color and the monochrome TV receiver is the portion including the decoder, color synch, and Y amplifier. While each of these sections will be discussed thoroughly in separate chapters later, the principles and main functions of each section should be understood now. Figure 4.5 shows a more detailed block diagram of the decoder sections and its associated networks. In general the decoder section consists of two different circuits. One portion of the decoder section removes the color sub-carrier and produces the two video signals, R-Y, B-Y, or the X and Z signals. Refer to Figure 3.9 for the X and Z vectors. This part is the demodulator referred to in Figure 4.5 and consists of a special type of detector which is synchronized by the color reference signal, the 3.579 mHz sine wave obtained from the color synch section. Since the X and Z signals were modulated onto the sub-carriers with a 62 degree phase difference, the two reference signals used to demodulate them are also 62 degrees apart in phase as shown in Figure 4.5.

From the demodulator emerge two video signals, the X and the Z signal or the R-Y, B-Y signals. In most receivers the —X and —Z signals are obtained simply by taking one polarity from the plate and the opposite polarity from the cathode of the demodulator tube. As was discussed in Chapter 3, the electrical conversion of the X and Z signals to the color difference signals is a simple matter of voltage addition and subtraction, taking phase angles into account. Usually additional gain is needed and separate amplifiers are used to bring each of the three color difference signals to the respective electron gun in the color picture tube. The color difference signals are converted into the actual color signals by the addition of the Y, the brightness signal and this addition takes place in the electron guns of the color picture tube itself. Figure 4.5 shows the paths of the various signals and provides a clear understanding of these signals and their transformations. The full details of demodulating and matrixing networks are found in Chapter 9.

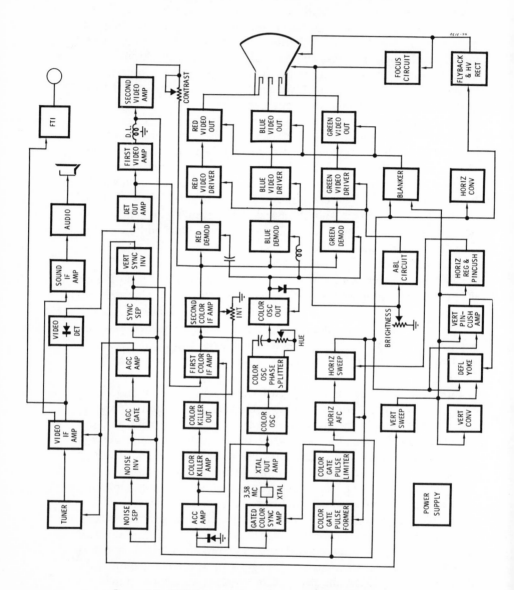

Figure 4.6—DETAILED COLOR RECEIVER BLOCK DIAGRAM

Returning to the overall operation of a color TV receiver, Figure 4.6 shows a detailed block diagram of a complete color set. For purposes of comparison a block diagram of a typical monochrome receiver is shown in Figure 4.7. Note that the color receiver is far more complex, but that a number of sections seem to be identical in both receivers. In actual circuitry even these sections, with the exception of the audio amplifier and FM detector, are not quite the same for both receivers. For example, the power supply for the color set must supply much more power than its monochrome counterpart because the color set has many more components. Similarly, the IF section in the color set contains more tuned circuits and has a much more critical alignment than the monochrome IF because cross-modulation of the color sub-carrier and the sound IF as well as phase shift of the color side bands must be avoided.

Even the RF tuner on the color receiver is more critical since excessive oscillator drift may cause a shift in the location of the color sub-carrier on the response curve. This in turn would cause phase shift of some of the color information and would result in wrong colors on the screen.

Color Reproduction

All of the sections shown in the block diagram of Figure 4.6 have one purpose—to make colored pictures appear on the screen. The HV section provides the power for the electron beams, the horizontal and vertical sweep sections sweep the electron beams over the screen area and the three video signals control the luminance of their respective colors as they light up on the screen.

In the next chapter there will be a detailed discussion of color picture tubes, a description of each of the different types in use, and many operational and circuit data. Before considering color picture tubes, however, we must understand the problems of reproducing a colored picture. A good example of color reproduction is the printing process used for the color illustrations in this book. Close observation will reveal that a color picture consists of small elements of the three or four basic colors. In color TV, three primary colors are used and the transmitted picture consists of small elements of these three colors. One of the salient features of the color TV system is the fact that all three colors appear simultaneously on the screen. This means that it is possible to have three separate picture tubes, one producing a red, one a green, and one a blue picture. When these three images are combined, optically, the proper full color picture will result. Some of the early color receivers employed such a three-tube color reproducer. Just

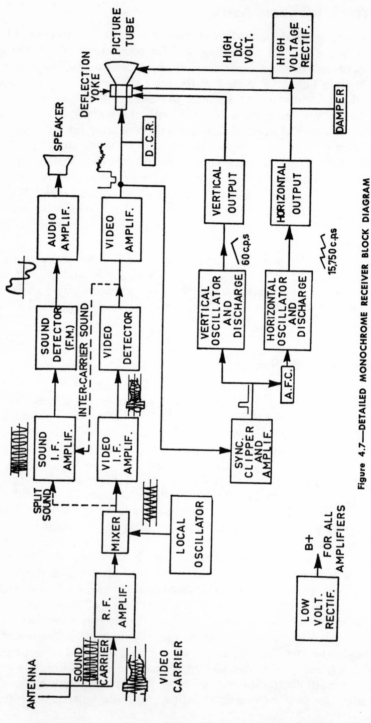

Figure 4.7—DETAILED MONOCHROME RECEIVER BLOCK DIAGRAM

42

as the three color cameras dissect the scene into three portions at the studio, so does the color picture tube reproduce the three portions again.

Unlike the three tube reproducer, the color picture tube does not need a special optical system to combine the three colors. Instead, the color tube relies on the integrating quality of the eye to combine the colors, just as the color printing process does. From a short distance the individual color elements in the color illustrations are not discernible, but a full color picture appears instead. In a similar manner the fine dots of the color picture tube screen add up to a complete color picture.

Since the brightness of these dots is determined by the red, green, and blue video signals it is imperative that these signals be exactly in step with each other. If, for example, the blue video signal were slightly delayed from the other two signals, a particular picture element which should appear as white would appear as yellow instead, while the subsequent colors would contain too much blue. It is essential that the time and phase relationships as well as the amplitude relationship of all three video signals be exactly the same as they were at the output of the three color cameras at the studio. For this reason it was necessary to provide a phase delay equalization network at the transmitter and a delay network is also used in the receiver to assure proper phase coincidence of all three signals.

Phase delay and phase relationships in general play no large part in monochrome TV, but they are of prime importance in color receivers. In addition to frequency response characteristics we must also know and control the phase delay in the various video amplifiers. Chapter 9 contains more details on phase delay networks.

As stated above, the color picture tube is the heart of the color receiver, and all other sections operate to supply the proper signals to it. Since three primary colors must be reproduced, all color picture tubes must have a screen capable of lighting up in all three colors. A variety of color tubes are now in use and many more types are under development. The next chapter contains a detailed discussion of the color picture tubes, their principles, accessories, and the circuits used with each type.

5

COLOR
PICTURE TUBES

Three-Color Reproduction

In Chapter 2 the principles of colorimetry illustrated that it is possible to reproduce a colored picture by means of three colored light sources, especially if the three colors are the primary colors. The three primary colors for TV, red, green, and blue, can be considered as three axes in the color pyramid and therefore will reproduce any of the colors enclosed in the color space of these axes. Reducing this to a practical example, as shown in Figure 5.1, three lights could project

Figure 5.1—THREE-LIGHT COLOR REPRODUCER

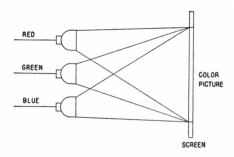

RED

GREEN

BLUE

COLOR
PICTURE

SCREEN

three color slides, each containing only red, green, or blue respectively, and on the white viewing screen a natural color picture would appear. This is a feasible demonstration of color adding; its only limitation is the fact that only still pictures can be projected. Needless to say that the three projected pictures must be in perfect register.

The example of Figure 5.1 represents one type of three color reproduction system, suitable even for TV if the three slide projectors are replaced with TV projection tubes, each having a colored filter over the screen. In fact, at the beginning of color TV several such receivers were demonstrated before the FCC and the public. Disadvantages of this system include relatively low brightness, difficulty of maintaining perfect register of three tubes and the associated optical system and the high cost of all this elaborate equipment. To overcome some of the objections of the three color projection system, three separate direct view tubes were used, each having a colored filter in front. The three pictures were combined optically by means of a set of dichroic mirrors. A dichroic mirror has the property of passing all light except that of one particular color. In the simplified dichroic display of Figure 5.2 mirrors A and B are at right angles and intersect each other. Mirror A is transparent except for red light and mirror B is transparent except for blue. As a result the green image will be seen directly, the red due to reflection from mirror A and the blue picture appears reflected on mirror B. The viewer therefore sees all three pictures superimposed, which gives the effect of a full colored scene.

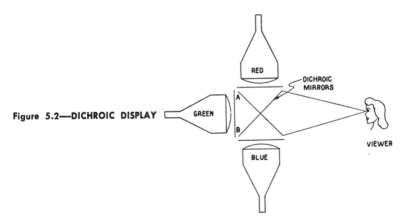

Figure 5.2—DICHROIC DISPLAY

The next step from a color reproducing system of three picture tubes is a single picture tube having three colored screens and some system of combining the three colors. This is found in the basic principles of the current shadow mask, dot screen picture tubes.

Phosphor dot screen. In monochrome picture tubes the exact color of the screen depends on the chemical mixture of the phosphor. Oscilloscope cathode ray tubes have either a green or blue color and, in some, red glowing phosphors are also available. In the color picture tube, red, green, and blue colored phosphor is used. This, when struck by the electron beam, lights up in its respective colors.

Various methods of incorporating three colors in a single screen have been tried out. Some of the experimental picture tubes contained three layers of phosphors, each activated by different velocity electron beams. Another type had a screen made up of many tiny three-sided pyramids, one side for each colored phosphor. The screen was activated by three electron beams, each coming from a different direction. Still other systems of depositing the three phosphors were tried out and finally only the dot method reached commercial acceptance.

Returning to the example of color printing—which is essentially a dot system—consider a picture tube screen where dots of three colored phosphors are deposited as in Figure 5.3. If the dots are small enough,

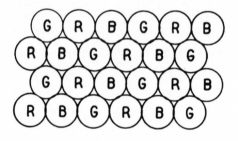

Figure 5.3—COLOR DOT ARRANGEMENT

they will be indistinguishable as individual dots and only the overall pattern will be visible, just as in printed color pictures. Thus the inability of the human eye to distinguish very small colored areas is used to combine the three colored screens. By intermixing the three screens in the form of a dot pattern the eye is made to perform the same function as the projection system and the dichroic mirror in Figures 5.1 and 5.2 respectively.

It is quite possible to excite each of the three colored dots in turn, by the same electron beam, but in order to maintain the principle of simultaneous color presentation three separate electron beams are used.

Each of these beams is supposed to strike only dots of its respective color, simulating the effect of three separate picture tubes. The fact that all three electron beams are inside the same tube envelope simplifies the registration problem greatly. Now all three beams are deflected by the same deflection yoke and can be centered together on the tricolor screen.

Basic Shadow Mask Tube

Many present day color TV picture tubes operate on certain identical principles and vary only in picture tube size, shape, and the deflection angle. These picture tubes use three separate electron guns which are usually arranged in an equilateral triangle. As seen from the rear socket of the picture tube, each of the three electron guns is located on a circle and spaced 120 degrees from its neighbor. All electron guns contain the same basic arrangements of filament, cathode control grids, screen grids, and focusing element. In metal envelope types the bulb of the picture tube forms the second anode or ultor, and in glass picture tubes the inside of the glass is coated just as in the black and white equivalents. The shadow mask picture tube derives its name from the inter-position of a large mask with holes which the three electron beams must pass as illustrated in Figure 5.4. To assure that each electron beam will only illuminate its respective color dot the shadow mask contains

Figure 5.4—THREE BEAMS CROSSING TO HIT SCREEN

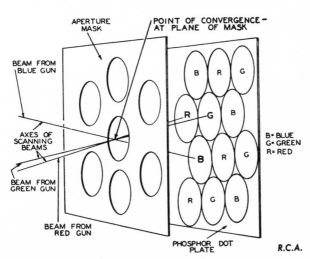

one hole for every group of three color dots on the screen. The construction and assembly of the shadow mask and color dot screen assembly is the most critical part in the manufacture of the shadow mask
picture tubes and probably accounts for most of the cost. It is apparent
from Figure 5.4 that each of the holes in the shadow mask must line
up very exactly with the three color dots and with the three electron
guns as well. Later paragraphs will illustrate the various magnetic and
electronic devices necessary to assure that the three electron beams
converge at exactly the right hole and strike the three color dots correctly. The brightness of each of the three phosphor dots depends on
the beam current provided by its respective electron gun. This current,
in turn, is the result of the cathode to grid potential and of the voltage
on the screen grid. The second anode or ultor is common to all three
electron beams. Similarly, the focusing elements, while mechanically
separated, are electrically connected inside the picture tube and a single
focus control provides the focus for all three electron beams.

Figure 5.5—CRT—COMPLETE ASSEMBLY

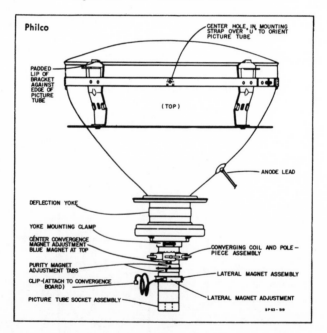

When a monochrome picture is received and the three cathodes are connected together, with the control grids all at the same potential, the monochrome picture will cause equal beam currents from all three electron guns. In order to produce a truly white, or neutral color picture, it is necessary that the three screen grids be set at different voltages so that the sum of the red, green and blue light emanating from the phosphor dots produces the effect of white. As will be shown in a later chapter, the adjustment for neutral grey shading does indeed depend on the setting of the three screen grid controls.

Figure 5.5 shows the external components mounted on a typical color picture tube. In addition to the mechanical strap assembly, there is a conventional high voltage connection to the second anode and then comes the deflection yoke. The deflection yoke, basically the same as for black and white tubes, must provide much more linear deflection in both the vertical and horizontal direction. All color picture tubes use electrostatic focusing and therefore no external focusing elements or devices are required. Instead, three new external mechanical assemblies appear. The first is the convergence assembly. Following that is the purity magnetic assembly, and closest to the picture tube socket, is the blue lateral magnetic assembly. The basic function of all three assemblies is to assure that each of the electron beams will only strike its correct phosphor dots on the screen.

Purity Assembly

The purity magnet affects all three electron beams to approximately the same extent. In order to make the adjustment, however, two of the electron beams, usually the blue and the green, are turned off and only the red dots are illuminated. When the color purity is correct, nothing but red of a single shade will appear on the screen. Figure 5.6 shows the action of the purity magnet, a circular magnetic ring with two adjustment tabs, together with a very enlarged view of the phosphor dots. When the entire purity magnet is rotated, the electron beam within the three electron guns is rotated as indicated by the circular arrows. When the two adjustment tabs are spread apart this increases the intentisity of the magnetic field and moves the electron beams in the radial direction. In most shadow mask tubes it is difficult to obtain perfect purity over the entire screen and a compromise setting is usually required so that the maximum area, possibly with the exception of the few edges, is pure red. To make sure of this, a low power microscope is often used to look at the dots themselves and verify that only the red ones are illuminated. Even a small glow of the green dots, for example, will give the impression of an orange rather than a red area.

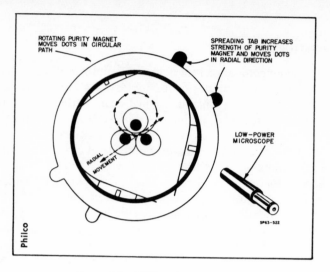

ROTATING PURITY MAGNET
MOVES DOTS IN CIRCULAR
PATH

SPREADING TAB INCREASES
STRENGTH OF PURITY
MAGNET AND MOVES DOTS
IN RADIAL DIRECTION

LOW–POWER
MICROSCOPE

RADIAL
MOVEMENT

Philco

SP63–522

Figure 5.6—PURITY MAGNET ACTION

The purity of the color display is affected by any stray magnetic fields and many such fields, unfortunately, exist. Moving a color TV receiver across the magnetic flux lines of the earth's magnetic field will cause certain parts of the chassis and the picture tube to become magnetized and this will result in spot impurities on the screen. To avoid this it may be necessary to degauss the entire color picture tube. Degaussing a picture tube, just like demagnetizing the tape recorder head, is accomplished by generating a strong AC field around the degaussed object and then gradually reducing this field. In many color TV receivers a degaussing coil is built in and usually located around the screen. When the TV set is first turned on, some current passes through this coil, usually a portion of the unrectified AC line voltage, and this is then gradually reduced as the set warms up. This method automatically degausses the picture tube every time the set is turned on.

Convergence Assembly

To make sure that the three electron beams reach the correct hole in the shadow mask tube, at the proper angle so that they will strike their respective color dots, a complex assembly, the convergence assembly of Figure 5.7 is used. The principle of the magnetic convergence system is shown in Figure 5.8 in the form of a cross section of the neck of the color picture tube. Each of the three electron guns, must be aligned in the correct radial direction, as indicated by the arrows of Figure 5.8. This is accomplished by means of internal pole pieces which are coupled, through the glass envelope, to magnetic external pole pieces. In Figure 5.8 we have shown a simple coil going

50

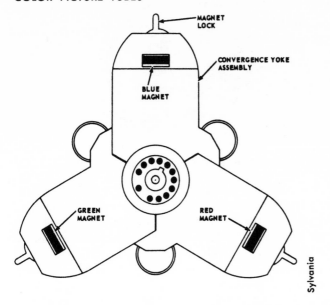

Figure 5.7—CONVERGENCE ASSEMBLY—REAR VIEW

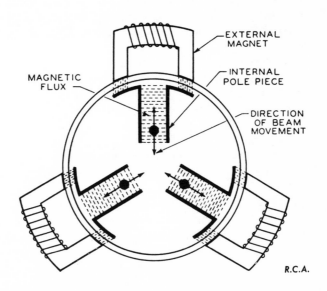

Figure 5.8—MAGNET CONVERGENCE PRINCIPLES

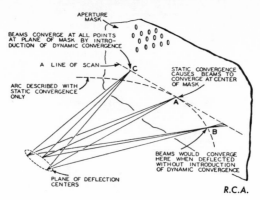

APERTURE MASK

BEAMS CONVERGE AT ALL POINTS
AT PLANE OF MASK BY INTRO-
DUCTION OF DYNAMIC CONVERGENCE

A LINE OF SCAN

ARC DESCRIBED WITH
STATIC CONVERGENCE
ONLY

STATIC CONVERGENCE
CAUSES BEAMS TO
CONVERGE AT CENTER
OF MASK

C

A

B

BEAMS WOULD CONVERGE
HERE WHEN DEFLECTED
WITHOUT INTRODUCTION
OF DYNAMIC CONVERGENCE

PLANE OF DEFLECTION
CENTERS

R.C.A.

Figure 5.9—ARC MOTION OF BEAM, NEED FOR DYNAMIC CORRECTION

around each of the pole pieces. If the current through the coil is changed, the flux across the pole piece changes and this will move the beam in and out radially as shown by the arrows. Because of the internal pole pieces, the action of each external magnet is confined to its respective beam, but we can still perceive two limitations of this system. The first limitation becomes apparent when looking at Figure 5.9.

If three electron beams are converged correctly at the center of the picture tube it becomes apparent that, as they describe an arc across the screen they will go out of convergence unless both the shadow mask and the screen are semi-circular. The screen is practically flat and the shadow mask behind it is also flat. What is needed here is a means of moving the point of convergence, B and C in Figure 5.9, as the electron beams sweep across the screen. This can be accomplished by passing a varying current through the coils of the external magnetic convergence assembly. The circuits generating these currents and the actual wave shapes and their controls are discussed in more detail in the next chapter. In essence, however, they must be roughly semi-circular or parabolic in shape to compensate for the difference between the normal arc traced out by the deflected electronic beam and the actual shape of the shadow mask and the screen.

The second limitation of the convergence principle shown in Figure 5.8 deals with the movement of the three electron beams towards the center.

In Figure 5.10 the blue lateral position magnetic intensity is shown as controlled by current through the coil. Actually, recent color TV receivers use a permanent magnet in an assembly to adjust the magnetic flux strength and thereby the lateral movement of a blue electron beam.

To allow some lateral movement of one of the beams an additional magnetic field is introduced. Its principles are shown in Figure 5.10.

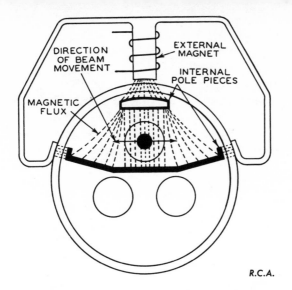

Figure 5.10—BLUE POSITION MAGNET PRINCIPLE

This blue positioning magnet again operates in a defined field which affects only the blue electron beam. Again the magnetic flux passes through the glass envelope but this time a variation in magnet strength causes only lateral movement. The advantage of this system lies in the complete independence of this adjustment with other settings.

Many recent color TV receivers, particularly those having screen sizes of 17 inches or smaller, use a new type of picture tube in which the convergence and purity adjustments are greatly simplified. The major difference between them and the original tri-color dot shadow mask

Figure 5.11—R.C.A. PRECISION IN-LINE COLOR TUBE

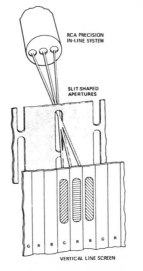

RCA PRECISION
IN-LINE SYSTEM

SLIT-SHAPED
APERTURES

G R B G R B G R

VERTICAL LINE SCREEN

53

tube is a phosphor screen made up of vertical stripes instead of dots and an in-line arrangement of the electron guns instead of a three-cornered or delta system. As illustrated in Figure 5.11, the R.C.A. in-line color tube uses an electrode gun structure in which the red, green, and blue electron guns are arranged next to each other in one line. The screen

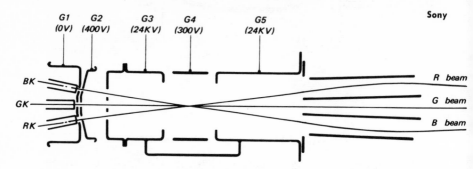

Figure 5.12—CROSS SECTION OF TRINITRON COLOR TUBE

consists of thin vertical red, green, and blue stripes, and instead of round holes, the mask behind the screen uses slit-shaped apertures. The first company to produce this basic arrangement was the Japanese Sony, which used an aperture grill consisting of vertical bars or wires. In 1965 G.E. produced a small screen in-line picture tube, and since 1973 all of the major manufacturers have turned to the in-line electron gun and vertical stripe screen approach.

The main advantage of the in-line system is the greatly simplified convergence adjustment. In the case of the Sony Trinitron color picture tube, there are only two separate convergence adjustments, which is a great improvement over the 12 adjustments required for the old shadow-mask, color dot type of picture tube. Another Japanese product, the Linytron, used in Sharp and Toshiba color TV receivers, also uses a two-control convergence arrangement. In the R.C.A. in-line picture tube, all of the convergence adustments are factory preset and located in the deflection yoke assembly which is rigidly cemented to the picture tube. This means that the deflection yoke assembly is replaced together with the tube and no convergence adjustments at all are required by the service technician.

Figure 5.12 shows a cross section of the Trinitron color tube system. Three separate cathodes, one for each color, are controlled by a single control grid, G1, and a single acceleration anode, G2. Focusing for all three beams is accomplished by means of G3, G4, and G5, a typical high voltage focusing arrangement. After emerging from the anode, labelled here as G5, each of the three electron beams travels in a separate path toward the grid structure which is located behind the vertical stripes of the screen material. In the Trinitron system, the small bridges shown as part of the slit-shaped aperture structure of Figure 5.11 are not present, and this results in slightly more brightness. The cathode structures shown in Figure 5.12 are angled towards the center of the focal point, but in the R.C.A. electron gun structure, the cathodes, as well as grids 1 and 2, are perfectly parallel and depend on the design of the deflection yoke to achieve the correction for the curvature of the screen.

Another recent improvement for all types of color picture tubes, whether they use a dot or vertical line screen, is the use of the so-called "black matrix" effect. This means that the spaces between color dots or color stripes are filled in with a black, light-absorbing material. The major advantage of this technique is to reduce glare and provide more clearly defined color elements. This technique is also sometimes called "negative guard band principle."

All of the improvements found in recent monochrome picture tubes have also been incorporated in color tubes. More efficient cathode construction and materials, a better vacuum sealing technique and, finally, the integral face plate arrangement. By the latter we mean the use of a safety glass which is permanently bonded to the picture tube face plate so that a separate safety glass is not needed. The bonded face plate is usually slightly etched to reduce reflections and generally provides a more pleasing appearance, both when the set is turned off and when a picture is displayed.

Metal envelope color picture tubes are also on the market and these differ from their all glass cousins only by the fact that the glass bulb is replaced by a metal envelope. Glass to metal seals bond the glass screen assembly and the glass neck to the metal envelope. Just as in monochrome receivers, the metal envelope acts as second anode and is connected to the high voltage source. A plastic insulating sleeve covers most of the metal envelope and suitable insulated mounting surfaces are provided.

COLOR
PICTURE TUBE CIRCUITS

To produce a proper color picture on the screen of a color picture tube, three types of circuits and components are required: DC operating circuits, AC signal circuits, and external correction circuits. The DC operating voltages are essentially the same as in monochrome picture tubes but must supply three electron guns. AC signals account for variations in brightness and color and control the three electron guns to reproduce the transmitted scenes. In the third group, we describe those external components and circuits which are necessary to converge the three electron beams through their correct holes in the shadow mask and to maintain correct colors all over the screen.

DC Operating Circuits

As in any cathode ray tube, the current in the electron beam is determined by the cathode-grid bias and by the potential on the screen grid. These voltages are adjustable to various levels of brightness. The second anode voltage is generally fixed, but some adjustment is provided to assure correct focus. As illustrated in the simplified circuit diagram of Figure 6.1, the basic adjustments and voltage relationships are the same for each of the three electron guns and are really no different than for monochrome picture tubes. Second anode potentials for color picture tubes range from about 22 kilovolts to 27 kilovolts, depending on the screen size and picture tube type. Electrostatic high voltage focus is used in most color picture tubes and this requires

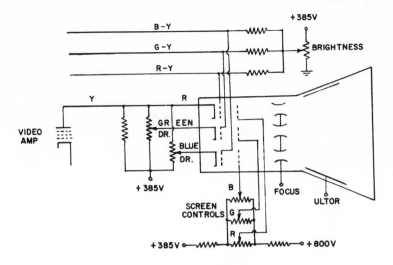

Figure 6.1—OPERATING VOLTAGES FOR COLOR CRT

voltages between 3 and 5 kilovolts. Note that all three focus elements are connected together and a single adjustment provides focus for all three electron beams.

In Figure 6.1 each of the three screen grids is brought out to a separate potentiometer which is connected between 800 volts obtained from the boost voltage in the flyback section and the 385 volts B+. The setting of the three controls determines the relative color balance since, as we know from colorimetry, the relative amounts of red, green and blue will determine the shading of the raster. This adjustment will be discussed in more detail in Chapter 15.

As illustrated in Figure 6.1, the three control grids for the blue, red and green electron guns are all at the same DC potential. If their cathodes were also at a common DC potential, the luminance of each of the three colors would be determined only by the screen grid controls. The actual DC potential between the three cathodes and the three grids is controlled by the brightness adjustment, a front panel control, which varies the DC setting for all three control grids as illustrated in Figure 6.1. In some receivers the control grids are returned to a fixed DC voltage and the bias between grids and cathodes is varied at the cathodes. In some receivers this is accomplished by varying the DC current through the video amplifer, either by putting the brightness control into the cathode of the video amplifier, or by adjusting the DC grid bias on the video amplifier. In either circuit, the DC potential between control grids and the three respective cathodes is set for a certain level of overall brightness. The three screen grid controls can adjust the relative brightness of each of the three primary

57

colors and thereby the color of the screen. When the three screen grids are properly adjusted and no video signal is received, the picture tube screen should appear a dim neutral grey. A few color sets have individual potentiometers to adjust the bias between each cathode and its control grid. This eliminates the need for screen grid adjustments, but complicates the color set-up procedure.

Although not shown in Figure 6.1, each of the electron guns has its own filament winding. In most picture tubes it is possible to observe the filament of each of the three electron guns and thus make sure that all three guns receive filament voltage.

In Figure 6.1 we have shown the green and blue drive adjustment potentiometers. While these adjustments have some effect on the DC operation, their main purpose is to balance out the red, green and blue video components as they are provided by the video amplifier. The adjustment of these controls will be discussed in detail in Chapter 15, but for the moment these controls can be neglected since their primary purpose is the adjustment of AC signals. The high voltage electrostatic focus used in practically all modern color picture tubes ranges from approximately 3 to 7 kilovolts. It is obtained from the flyback section, together with the 20 to 27 kilovolts ultor voltage.

HV Power Supplies

Just as in monochrome TV receivers, the high voltage required for the second anode of the picture tube is obtained from the horizontal flyback transformer. In color TV sets, however, two additional requirements must be met. The high voltage for the second anode must provide enough current for three electron beams and it must be much better regulated than in the monochrome picture tube. The focusing high voltage must have some adjustment range and can be obtained in a variety of ways. In some receivers the focusing potential is derived from the high voltage for the second anode by means of a bleeder network. In other receivers a separate high voltage rectifier is used, employing either a vacuum tube or else a silicon diode assembly. The focus adjustment can be a potentiometer or, in some of the more recent receivers, an inductor is tuned to control the amount of pulse power which is rectified. Regardless of the refinements, however, the focusing voltage does not require regulation and, once it is adjusted for good focus, further adjustments are usually not required.

The second anode voltage, however, must be set pretty accurately and the voltage must remain constant, regardless of current fluctuations. The basic reason for this is that the second anode serves three electron guns. Assume that the red gun draws a substantial amount of

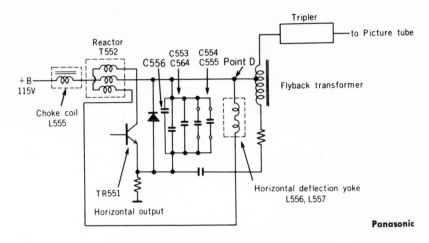

Figure 6.2—HV REGULATOR CIRCUIT

current, due to a large bright red area on the screen. If the anode voltage is now reduced, the blue and green electron beams will operate at a lower anode potential and this will reduce their brightness. Certain portions of the picture may have the wrong colors because the blue and green components are wrong. The current drain due to one or two of the electron guns must not affect the voltage that is available to the remaining electron beams. For this reason regulation of the second anode voltage is a requirement and is provided in all color TV receivers.

The high voltage regulator circuit of Figure 6.2 is typical of many color TV receivers and illustrates how the high voltage for the color picture tube is adjusted and automatically regulated. DC power is supplied to the flyback transformer through the primary winding of the saturable reactor T552. The secondary windings of the saturable reactor are connected in parallel with the horizontal deflection coil, and one end of the secondary goes to the collector of the horizontal output transistor TR551. The width and amplitude of the flyback pulse at point D, on the flyback transformer, depend on the inductance of the two deflection coils, on the saturable reactor T552, and on the array of parallel capacitors, C553, C554, etc. When the picture tube current increases, the DC current through the saturable reactor, T552, increases as well. An increase in DC current reduces the inductance of the reactor by increasing the flux of the core. This causes a decrease in the pulse width and an increase in pulse amplitude. Because the pulse amplitude increases, the high voltage at the picture tube does not decrease even though there is more current drawn.

As shown in Figure 6.2, two of the capacitors, C554 and C555, can be connected into the circuit by means of a simple strapping arrangement. To increase the high voltage at the picture tube one or both of these capacitors are disconnected, and to decrease high voltage, one or both of them are connected in the circuit. This does not permit a very accurate adjustment, but because all of the components have been selected to provide the required high voltage in the first place, only a small amount of adjustment is ever needed.

The rest of the high voltage circuit contains elements with which the reader should be familiar. The silicon diode shown next to C556 is the damper diode, and the high voltage rectifier in this particular circuit is a voltage tripler, a circuit that will be discussed in more detail in Chapter 11.

In many of the most recent color TV receivers the high voltage regulation is performed in the output amplifier of the horizontal flyback section. These circuits control not only the high voltage but also the horizontal drive in an automatic, feedback system. Some of the circuits used for the high voltage regulation will be described as part of the horizontal flyback system in Chapter 11.

The reader familiar with monochrome television servicing need not be reminded of the hazards of the high voltage section. Color TV receivers use even higher voltages and somewhat larger currents and all of the precautions prescribed for monochrome receivers apply even more stringently to color. Arcing and Corona are somewhat more of a problem in color TV receivers because of the higher voltages. Recent design improvements, in the form of plastic sleeving, molded insulation, etc. have reduced this problem.

AC Signals

In addition to the DC voltages illustrated in Figure 6.1, the color picture tube must, of course, be supplied with the AC signals necessary to modulate the electron beams. In monochrome TV receivers the AC or video signal was applied either to the grid or cathode, with the other element remaining at a fixed DC potential. In many color TV receivers, each electron gun also performs the addition of the brightness signal and color difference signal. This means that, if the brightness signal [Y] is applied at the cathode, the color difference signal [R-Y, G-Y, B-Y] is applied on the control grid. Since the same brightness signal is added to all three color difference signals, the three cathodes can be connected together, as illustrated in Figure 6.1. The three control grids receive their signals from the output of the matrixing amplifier circuit.

When a monochrome transmission is received on a color TV receiver, there will be no AC signal on the three control grids, but they will all be at the same DC potential. The three cathodes will all receive the brightness [Y] signal resulting in a monochrome picture. Each of the three electron guns generates one of the three primary colors on the screen and we have shown in Chapter 2 that different amounts of red, green and blue are required to produce a neutral grey or white picture. It is for this reason that the green and blue drive potentiometers shown in Figure 6.1 are provided to control the amount of green and blue brightness [Y] signal relative to a fixed amount of red.

The brightness [Y] signal contains all of the picture detail and should therefore have a bandwidth close to 4 mHz. In addition, the brightness signal should be as linear as possible and should have a DC component corresponding to that of the original picture. While in monochrome TV receivers the bandwidth, the linearity, or the DC component can have considerable tolerance, in color TV receivers each of these three parameters is essential. The video amplifier section for the brightness channel in a typical color TV receiver is therefore quite elaborate.

Figure 6.3 shows a three-stage amplifier, with the delay line connected between the first and second stage. Among the number of interesting features of this circuit is the fact that there is direct coupling from the video detector to the output amplifier. One of the reasons why three transistors are used in this circuit is for impedance matching, as can be seen by the fact that the first stage and the last stage are both emitter followers. From the emitter of the first stage the 3.58 mHz color sub-carrier signal is taken off for amplification in the color IF section. Note that the base of Q1E contains a peaking coil and resistor to compensate for possible loss of high frequencies. Additional peaking coils are used in the last two stages. One portion of video signal is taken off at the base of Q2E and goes to the noise separator circuit. From the emitter of Q2E the opposite polarity signal goes to the synch separator and AGC section. The actual video signal that is applied to the picture tube has a DC component from 95 to 35 volts. A portion of its amplitude range is determined by the setting of the 250 ohm contrast control.

Earlier video amplifiers, particularly those using vacuum tubes, contained additional adjustments such as potentiometers to control the peaking of the high-frequency and the video signal. In those receivers, where separate potentiometers were used to set the cathode-grid bias for the picture tube, these three potentiometers, red, green, and blue, were also considered part of the video circuit. Most of the solid state

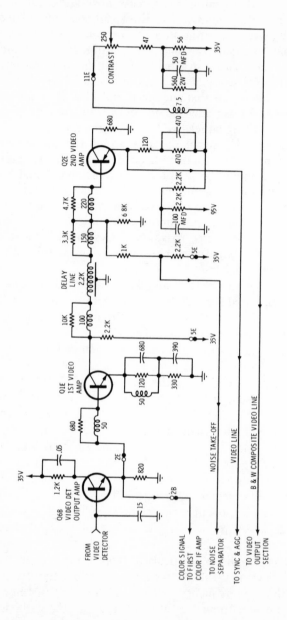

Motorola TS-915

Figure 6.3—VIDEO AMPLIFIER SECTION

62

color TV receivers manufactured since 1970, however, have eliminated these adjustments.

As in black and white TV receivers, the polarity of the video signal at the color picture tube must be such that the synch pulses cut off the picture tube. This means, if the video signal is applied to the cathode of the picture tube, the synch pulses must go positive. If it is applied to the picture tube grid, then the synch pulses must go negative.

Convergence Systems

In Chapter 5 we have seen the need for maintaining correct convergence over the entire screen. First we will describe the convergence system used for the dot type shadow mask tubes. The three electron beams are converged by external permanent magnets to pass through the holes of the shadow mask and strike the correct phosphor dots. We have seen the necessity for the lateral blue magnet in achieving correct convergence between the three electron beams. All of these adjustments, however, can only provide convergence in one relatively small area in the center of the screen. To compensate for the difference between the flat face plate and shadow mask, and the arc described by the electron beam, a correction voltage or current must be provided which will appear somewhat like the two curves shown in Figure 6.4. Early TV sets used electrostatic convergence with a high voltage signal applied on a special internal electrostatic convergence element. Practically all recent color TV receivers, however, use a magnetic convergence system with external electromagnets, controlled by a current which varies in step with the vertical and horizontal deflection signals. The horizontal and vertical correction currents are obtained directly from the respective deflection sections. In the case of the horizontal correction current,

Figure 6.4—HORIZONTAL AND VERTICAL CONVERGENCE SIGNALS

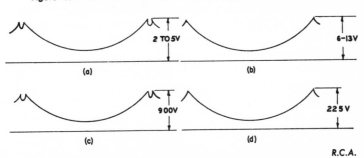

(a) 2 TO 5V (b) 6–13V

(c) 900V (d) 225V

R.C.A.

Figure 6.4a shows the slight dip at each end of the correction current which is due to the power required for the high voltage flyback section. Until the 1965 series of color TV receivers, it was common practice to combine the vertical and horizontal correction current in the same coils. A typical convergence circuit of this type, for one of the three electron guns, is shown in Figure 6.5.

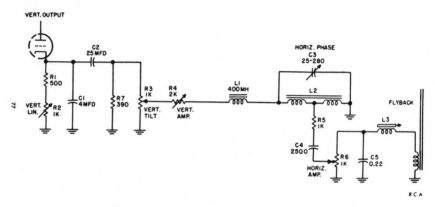

Figure 6.5—COMBINED CONVERGENCE SYSTEM

In this unit a small PM cylinder is inserted into each horseshoe-shaped core assembly and the position of this magnet determines the DC convergence. For vertical dynamic convergence a portion of the voltage across the vertical output amplifier cathode is used. This signal is applied across C1 and R7 and has a parabolic waveshape. R3 adjusts the tilt of the parabola and R4 its amplitude. L1 is simply a choke to keep the 15 kHz horizontal signal out of the vertical sweep section.

The horizontal dynamic convergence signal is obtained from a special winding on the flyback transformer. L3 and C5 determine the waveform of the horizontal signal. R6 serves as amplitude control while C3 tunes the resonant frequency slightly to permit phase shifting or tilt. L2 is the convergence coil itself, which forms a resonant circuit with C3. The purpose of R5 is to isolate each coil assembly and, in conjunction with C4, it helps keep the vertical signal out of the horizontal circuit. Although only one convergence coil circuit is shown in Figure 6.12 the other two are identical. The vertical signal for all three coil assemblies is derived across R7 and the horizontal signal across C5.

Since 1965 practically all color TV receivers use a new and simplified dynamic convergence system. The various chokes and capacitors used in the earlier circuit of Figure 6.5 are eliminated in the new system. In their place diodes are used for wave shaping and separate coils, wound on the same electromagnet, are used for the vertical and horizontal convergence signals. Another improvement, primarily of importance in the adjustment procedure, is the use of common controls for the green and the red convergence currents.

The blue convergence signals are adjusted by their own controls as illustrated in the simplified circuit for the vertical dynamic convergence shown in Figure 6.6. The blue convergence magnet receives the vertical correction current from the cathode of the vertical output amplifier through capacitor C1. The combined effect of R1, C2, R5 and the diode D1, together with the inductance of the convergence coils,

Figure 6.6—VERTICAL CONVERGENCE CIRCUIT

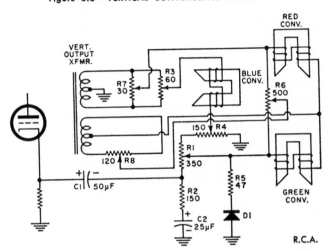

produces a parabolic wave shape. The amplitude of this current, as applied to the blue convergence magnet, is controlled by potentiometer R4. When this potentiometer is set close to the ground, very little current will flow through the windings of the blue convergence magnets. The wave shape of this current can be controlled by adding a small sawtooth current, obtained through the center tapped secondary of the vertical output transformer, and adjusted through resistor R3. When this resistor is set at its exact center, the opposing polarities of the sawtooth current will cancel out and the only dynamic control signal will be that coming from potentiometer R4. Setting R3 in either direction from the center will, in effect, add a small sawtooth signal to either side of the parabolic waveform. This gives the effect of a tilt, when the waveform is monitored on an oscilloscope. For this reason R3 is often referred to as the blue tilt control.

The red and green convergence magnets receive the same vertical correction, but its amplitude is controlled by R1. The tilt control for the red and green convergence is R7, connected in parallel with the blue convergence tilt control R3. It will be noted that the signal from R7 does not reach the red and green convergence coils in equal amounts because potentiometer R6 is connected between these two coils. Another tilt adjustment is provided by R8 which is connected across another winding of the vertical output transformer. The center tap of that winding goes to the common return of the red and green convergence coils. R8 will have an effect on the amount of tilt which is added to that provided by R7. The distribution of this added tilt signal is controlled by R6, which therefore can differentiate between the red and green convergence signal. We have seen that the blue convergence magnet is controlled by two potentiometers. One determines the amplitude of the symmetrical convergence signal, while the other determines the tilt of that signal. The red and green convergence magnets also have a total of 4 controls. Two of these determine the relative tilt or the relative convergence between red and green and two of them, R1 and R7, are used to set up the total convergence of red and green together. The six controls shown in Figure 6.6 are used to adjust the vertical convergence, top and bottom of the three electron beams.

The dynamic convergence circuit for the horizontal convergence coils is shown in Figure 6.7. A special winding on the flyback transformer provides the signal through C1 and T1 to the blue convergence magnet. Again the coil, together with R1 and diode D1 produces a parabolic waveshape. Because of the higher frequencies a tuned circuit, R2, C2 and L1, are used to regulate the amplitude. When L1 is tuned to resonance, it represents a higher impedance and more current flows through the convergence coil. R3 controls the time constant of C3-R1 and thereby affects the phase of the parabolic current as illustrated in Figure 6.7. The tuning of T1 also controls the phase, but it has its major effect on the other side of the waveform as illustrated. The blue horizontal convergence section has three controls: Amplitude, left tilt and right tilt. This is necessary since horizontal convergence must be provided over a wider range than vertical convergence.

The red and green convergence coils obtain their correction current from the same winding from the flyback transformer through C4 and L2. At L3 the signal is divided into two signals of opposite polarity and since they must be "in step" in the windings of the red and green convergence coils, one of the two is reversed in polarity. This has the

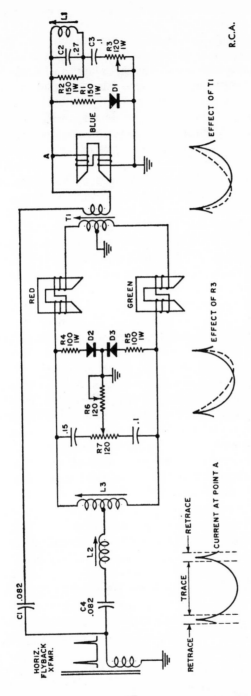

Figure 6.7—HORIZONTAL CONVERGENCE CIRCUIT

67

advantage that a minimum of interference can be expected in the magnet assembly itself. The resistor and diode combinations, together with the convergence coils and the center-tapped winding of T1 generate the parabolic waveshape. Note that L3 is in parallel with the combination of R7 and the two capacitors. L3 will shift the phase of the convergence current primarily in one direction while the R-C combination of R7 and R6 will shift it mostly in the opposite direction. R7 itself determines how much current goes to the red and green convergence coils respectively while R6 adjusts the actual phase. In a typical receiver L3 is used to adjust the convergence between the red and green electron beams at the right of the screen while R7 controls the convergence between red and green electron beams at the left. R6 will be adjusted to converge both red and green against the blue at the left of the screen while L3 controls both red and green at the right.

These designations, left and right, are arbitrary and it would be possible to reverse them. The manufacturer's choice of component values for these controls, however, provides the greatest range of adjustment for the areas of the screen indicated. As will be shown in some detail in Chapter 15, the actual adjustment instructions, test pattern and control markings on the convergence assembly itself makes the choice of these points practically automatic.

In the vertical circuit of Figure 6.6 we have seen a blue amplitude and a blue tilt control, R4 and R3 respectively. The red and green convergence are also controlled by four adjustments, but their function cannot be simply described as two for each color. In actual circuitry, the adjustment is also considerably simplified because, once the red—green difference adjustments are made, they usually do not vary very much and only the overall red—green controls are adjusted. The fact that the blue electron gun can be adjusted almost independently of the other two is also of great help.

In Chapter 5 we have described the great simplification in convergence possible by using in-line electron guns in the color picture tube. We have seen that in the R.C.A. version all of the convergence adjustments are performed at the factory. In the Sony color picture tubes and in similar in-line versions, the convergence system is only needed to shift electron beams in the same horizontal plane. The center beam is usually fixed, and only the left and right beams are adjusted in a sideways direction. In the Sony small-screen picture tubes, this correction is performed by a deflector element contained inside the picture tube element. This differs greatly from the electromagnetic convergence correction since it uses the electrostatic principle. In most receivers, the ratio of static, DC convergence voltage to the dynamic, parabola-shaped correction voltage is fixed. This means that only one control for the left and right electron means each performs all convergence adjustments.

Degaussing of the Color Picture Tube

In Chapter 5 the need for a purity magnet assembly and its basic operation was described along with the operation of the three static and dynamic convergence assemblies and the lateral blue magnetic adjustment. From all of this it becomes apparent, that a number of precisely controlled magnetic fields are necessary, to assure that each of the three electron beams reaches the shadow mask at the right angle and strikes the correct phosphor dots all over the screen. Unfortunately, many parts of the color television receiver, including the picture tube and its internal components, are made of ferrous materials and any of them can develop magnetic fields which might be in conflict with those specially created to provide clean, well converged color pictures. When a color TV receiver is shipped from one place to another, or even moved to various locations in the home, it passes through the different flux lines of the earth's magnetic field. Small metallic components in and around the picture tube can become magnetized by passing through the earth's magnetic field.

To overcome the effect of this stray magnetism, earlier TV picture tubes contained special permanent magnets, mounted in an assembly around the screen, which could be adjusted to correct for these spots. The more efficient approach to the stray magnetic field problem is to demagnetize, or degauss, the area of the color picture tube screen. This is accomplished by an AC magnetic field in a coil, approximately the size of the picture tube screen. When this magnetic field is gradually reduced, the area will be without any definite magnetic polarization. The degaussing coil is an important tool for technicians, particularly for those receivers in which a degaussing coil is not built in. Most TV models made since 1965 contain a built-in degaussing coil which is either permanently connected into the power supply of the receiver or else can be switched on whenever necessary.

The automatic degaussing circuit of Figure 6.8 is typical of the majority of recent TV receivers. The coil itself is wound in sections around the picture tube screen and is connected into the AC power supply, before the rectifiers. The operation of this circuit depends upon the characteristics of two resistors. The thermistor shown in Figure 6.8 is a resistor which has a relatively large resistance when cold and a very small resistance after it warms up. The voltage dependent resistor, VDR in Figure 6.8, acts in exactly the opposite way. When a large

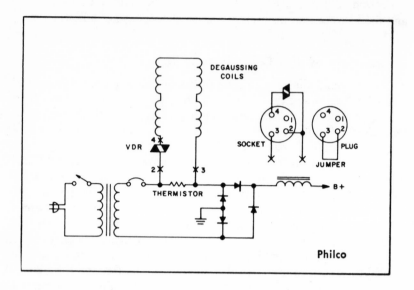

Figure 6.8—AUTOMATIC DEGAUSSING CIRCUIT

voltage across it decreases, its resistance becomes higher. When the color TV receiver is first turned on, the thermistor is cold and presents a large resistance, causing most of the 60 Hz current to flow through the VDR and the degaussing coil. As the thermistor warms up, less and less current passes through the degaussing coil and the voltage across the VDR increases. After the TV set has been on a minute or two, the VDR is very much larger than the thermistor, and most of the current will flow through the thermistor, with only a very small amount passing through the degaussing coils. This has the same effect as applying the line current to the coil and moving the coil away from the screen of the color picture tube. At first, a strong 60 Hz magnetic field is applied and as this field diminishes the demagnetization of any stray magnetic object in the vicinity is achieved.

The automatic degaussing scheme has the advantage that the color picture tube is demagnetized each time the set is turned on. In a few recent models, the degaussing coil is activated in a different manner. During receiver operation, a capacitor is kept charged up from the B+ boost voltage, and when the set is turned off, this capacitor is discharged through the degaussing coil. This causes a decaying current fluctuation through the degaussing coil which has the desired demagnetizing effect.

The major difference is, that degaussing occurs every time the receiver is turned off. A variety of automatic degaussing circuits are in use but most of them operate on the same principle as the power supply circuit shown in Figure 6.8.

7

A TYPICAL
COLOR TV RECEIVER

After studying the fundamentals of colorimetry, the color signal parameters and the color picture tube, the reader will be ready to look into the actual circuitry of a typical color TV receiver. Before, however, describing the detailed circuitry, the overall functions of the various circuit sections must be thoroughly understood.

The basic functions of the overall color TV receiver are shown in the simplified block diagram of Figure 7.1. The reader will readily recognize many of the sections which are the same in monochrome TV receivers. The VHF tuner, the UHF tuner, the video IF section and the entire audio section are certainly no different, at least at first glance, from those of monochrome sets. In the block diagram of Figure 7.1 we see that there are two video amplifier stages, something that is rare in late model monochrome sets. The synch and the AGC section, as well as the vertical and horizontal deflection portions are again the same as in monochrome receivers. Unique for color TV are the color IF section, often called the bandpass or chroma amplifier, the killer, the demodulators, the color synch and gate section, and, finally, the convergence section.

A more detailed block diagram of a typical color TV receiver is shown in Figure 7.2. The active components, such as transistors, diodes, and integrated circuits, are listed within each functional block. Beginning at the upper left-hand corner, we can see the UHF and the VHF tuner. Notice that the output of the UHF tuner is switched into the

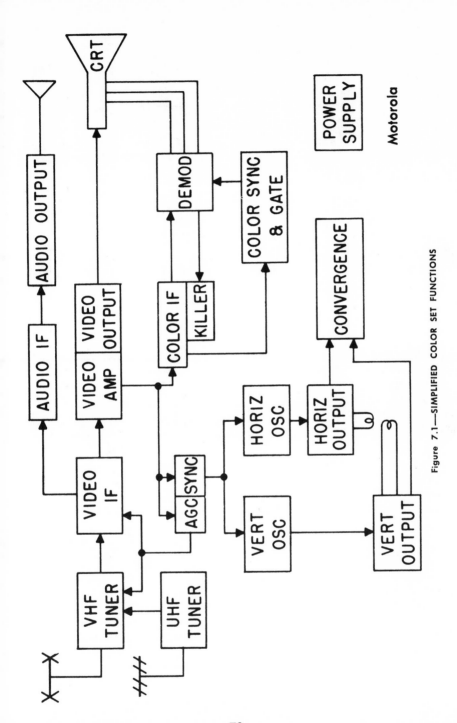

Figure 7.1—SIMPLIFIED COLOR SET FUNCTIONS

Motorola

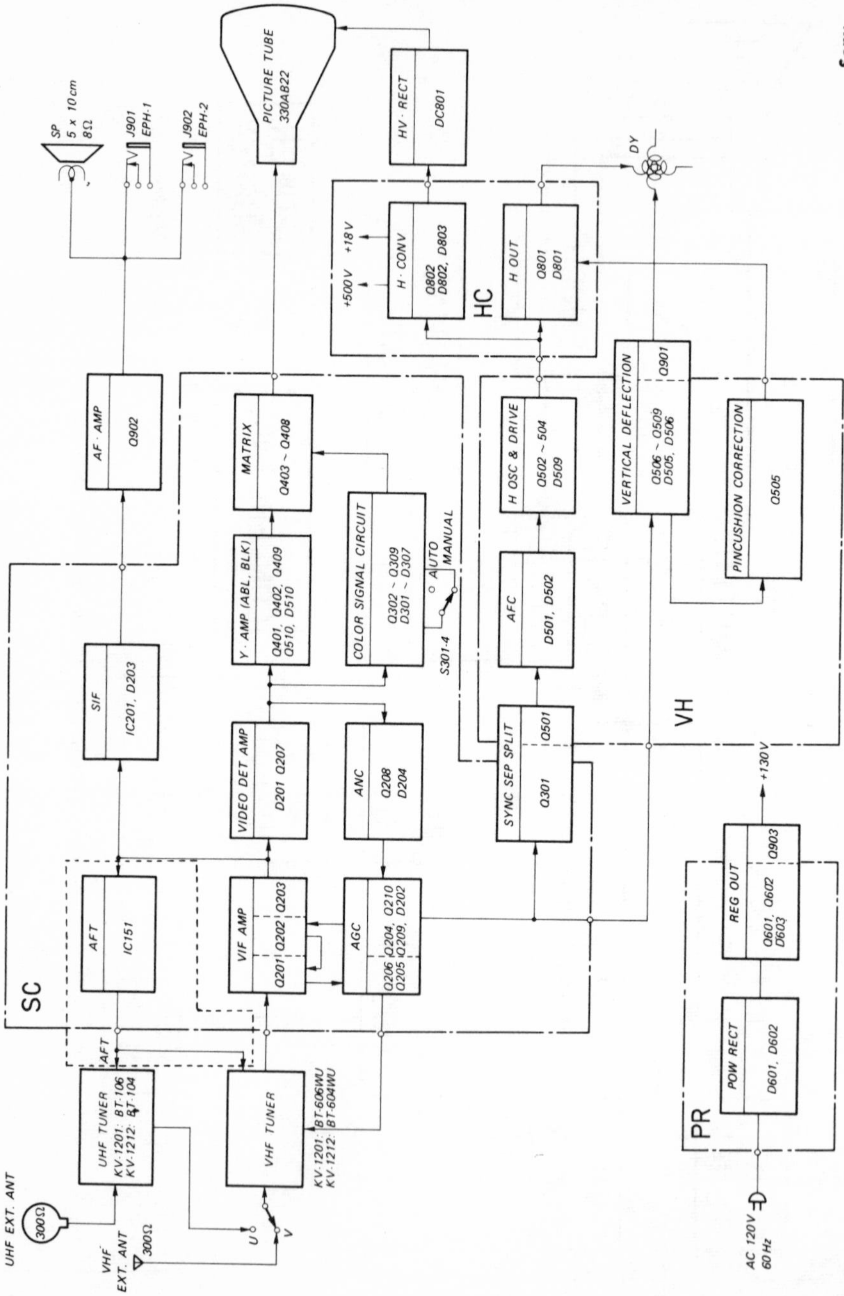

Figure 7.2—COLOR RECEIVER BLOCK DIAGRAM

Sony

74

antenna input of the VHF tuner, just as in monochrome TV receivers. The automatic fine tuning system (AFT) consists of a single integrated circuit—ICI51. The video IF amplifier contains three transistors, Q201 through 203. From the IF amplifier the signal goes to the AFT section and also to the sound IF section, where the 4.5 mHz intercarrier sound signal is obtained from the overall IF by the diode D203. The limiting and FM detecting function is performed by a single integrated circuit, IC 201. A single transistor audio amplifier, Q902, drives the speaker and the headphone jacks.

The video detector and amplifier section, D201 and Q207, provides video signals to the brightness amplifier and its associated circuits. In addition, the video signal goes to the automatic noise (ANC) control circuit which drives the automatic gain control (AGC) section. As in monochrome receivers, the AGC section controls the gain on the VHF tuner and the IF amplifier. The color signal circuit contains seven transistors and six diodes, and a more detailed discussion of the operation and function of these components is contained in Chapters 9 and 10. In essence, these circuits generate the color synchronization signals and perform the demodulation of the two color difference signals, which are then supplied to the matrix amplifier section. Here each of the color difference signals is combined with the brightness signal to drive the picture tube with the complete three-color video signals. The "Auto-Manual" switch refers to the automatic color control circuit which can be deactivated at the customer's option.

The remaining portion of Figure 7.2 does not differ very much from the functions normally expected in a monochrome TV receiver. The synch separator receives synch pulses from the automatic noise control circuit and provides the horizontal synchronizing pulses to the automatic frequency control circuit, a simple two-diode phase detector, which controls the horizontal oscillator. The horizontal output circuit provides the deflection voltage and also drives the horizontal convergence and high voltage for the picture tube. The only unique feature of the vertical deflection section is that it provides for pincushion correction. A regulated power supply is essential for any solid-state color TV receiver.

For the sake of simplicity, a number of connections and feedback paths have been omitted in the block diagram of Figure 7.2. The standard feedback from the horizontal output to the automatic frequency control, for example, has not been shown, and the color killer function, which depends on a portion of the flyback pulse, has also been omitted.

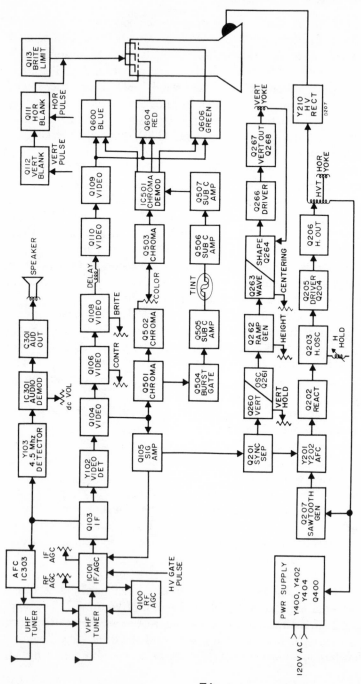

G.E.

Figure 7.3—DETAILED BLOCK DIAGRAM

76

A detailed, stage-by-stage block diagram is described in the next paragraphs.

Detailed Block Diagram

An explanation of the functions of every stage in a typical color TV receiver is contained in a detailed block diagram of Figure 7.3. As in the discussion of Figure 7.2, we start in the upper left-hand corner with the UHF and VHF tuner. In the receiver of Figure 7.3, the automatic fine tuning has been labelled AFC (automatic frequency control) instead of AFT, but its function is the same. The single IC101 provides IF amplification and AGC control, followed by the IF output stage, Q103. Note that in Figure 7.3 an adjustment potentiometer for the AGC bias for the RF stage and for the IF stage is shown separately. A separate RF AGC transistor, Q100, and the application of the high voltage gating pulse to the AGC circuit, is also shown in Figure 7.3. The audio circuit is essentially the same as that shown in Figure 7.2, but here the volume control is displayed as a separate function. From the video detector three stages of video amplification, including the contrast and brightness controls, deliver the brightness signal to the delay line. Two additional stages are used to drive the three-matrix and video output amplifiers.

The output of the first video amplifier is provided to a single transistor stage, Q105, which provides a feedback signal for the AGC circuit and which also drives the synch separator circuit. From the first video amplifier the signal goes through three stages which provide the 3.58 mHz color sub-carrier gain. Between the second and third stage is the color amplitude control. The first chroma or color amplifier also drives the burst gate which is controlled by the high voltage gating pulse, not shown in Figure 7.3 for simplicity's sake. In this particular receiver a crystal ringing circuit is used which generates a continuous reference signal from a single burst. For this reason, three sync signal amplifiers and a phase-shift type of tint control are shown here. In other receivers, as explained in detail in Chapter 10, different types of color synchronization circuits are used.

The output of the third sync signal amplifier and the output of the third chroma amplifier are applied to the demodulator portion which consists of a single integrated circuit, IC501. This circuit demodulates the red, green, and blue color difference signals from the chroma signal, based on the reference signal input. A detailed description of this type of circuit is contained in Chapter 9. The three color difference signals are applied to the three matrix amplifiers, Q600, Q604, and Q606, which then drive the respective control grids on the color picture tube. The cathode of the color picture tube receives the vertical and horizontal blanking signal as well as a DC brightness limiting voltage.

The detailed circuitry of the vertical and horizontal deflection portion shown in Figure 7.3, differs very little from its monochrome counterpart. Note that vertical hold, height, and centering, as well as horizontal hold controls, are shown here. The convergence assembly and its controls have been omitted from the block diagram of Figure 7.3.

The above discussion of color TV receiver block diagrams has shown the differences between them and their monochrome counterparts. It is clear that many of the circuit functions of color TV receivers are the same as those found in monochrome sets. A number of new functions have been introduced and they will be discussed in more detail in the following chapters. For the serviceman and troubleshooter, it is important to understand the different functions of each block in order to be able to isolate a defect to the correct receiver section.

Circuit Functions and Frequencies

In this section we shall consider only those circuits and frequencies that differ substantially from their monochrome counterparts. The first of these is the video IF where, in addition to the video and sound IF carrier, the color sub-carrier IF must also be considered. In the receivers described in the block diagrams in Figures 7.1, 7.2 and 7.3 the video IF is 45.75 mHz, the sound IF 41.25 mHz and the color sub-carrier IF is 42.17 mHz. Figure 7.4 shows the overall IF response required to properly amplify each of these frequencies. Note that the video IF carrier is located at the 50 per cent point just as in most monochrome IF response curves. A fairly flat top from 45 to 41.65 mHz is required here to pass the entire color sub-carrier and its side bands with minimum phase shift. Since the color sub-carrier is at 42.17 mHz, about 0.52 mHz of side band lies within the flat portion of the IF response curve. This includes all of the color sub-carrier side bands.

If the IF response curve shifts with respect to the color sub-carrier and its side bands, this will result in portions of the side band falling on the slope of the response curve. The phase shift of frequencies falling on the slope of the curve as well as their amplitude varies greatly from the phase shift and amplitude on the flat portion. In monochrome the only effect of such a shift is a slight degradation in picture detail, but in color TV serious picture defects will be introduced. When unequal phase shift and partial side band loss occurs in a color TV set this may become evident by the fact that flesh colors will be too red, a blue may appear as purple, yellow as orange, and so forth.

Occasionally it happens that the IF response curve is correct, but the colors appear quite wrong on one channel. This may be due to the fact that the local oscillator at that channel is mistuned or else drifts considerably after warm up, causing the apparent shift in IF response curve arrangement. As is pointed out in detail in Chapter 13, both the RF and IF section of a color TV receiver must be carefully aligned, first individually and then the overall alignment must be checked.

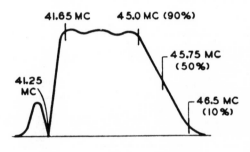

Figure 7.4—OVERALL IF RESPONSE

The response curve in Figure 7.4 indicates a sharp dip at the sound IF. This is required to avoid beat interference between the color sub-carrier and the sound carrier. We know that at the second detector the sound and video IF carrier beat to produce the 4.5 mHz intercarrier sound IF. Similarly the sound and the color sub-carrier can beat and then produce a 0.92 mHz interference signal. This would be visible as a fine beat pattern superimposed on the picture. To reduce the likelihood of this interference becoming apparent, the color sub-carrier is ordinarily transmitted at a lower level than the brightness signal and, in addition, the sound IF is attenuated greatly. In a typical receiver the sound IF is about 55 db below the flat portion of the response curve.

Circuitry of typical delay lines is described in detail in Chapter 11, but for a quick understanding let us consider a simple L C network, as in Figure 7.5. Any signal passing through this network will require some time to build up the magnetic field of the coil and to charge up the capacitor. If the values of L and C are chosen correctly a certain range of signal will be passed without effectively changing in appearance, but with some delay. For commercial delay lines, the arrangement shown in Figure 7.5 is repeated many times and the overall design permits fixed time delays to be specified for a certain frequency band. It should be pointed out that the amplitude at the output of the delay line is lower than at the input and every delay line has a certain power loss, however small. The time delay for the brightness channel is about 1.5 microseconds. This is the time required by the X and Z signals to be detected and appear at their respective video amplifiers. The delay line is a familiar component in radar and other pulse information systems.

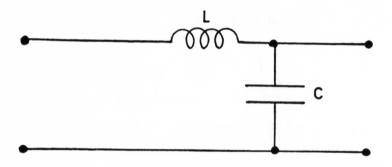

Figure 7.5—SIMPLE L-C NETWORK

The color sub-carrier amplifiers shown in the block diagrams of Figures 7.2 and 7.3 might have ideally a response curve which is different for the I and the Q signal. We have pointed out earlier, however, that most modern TV receivers do not use the broad bandwidth available for the I signal but use either the X and Z vectors or the R-Y and B-Y color difference vectors as reference for demodulation. For these demodulating systems the bandwidth normally ranges from about .5 to about .6 mHz above and below the sub-carrier.

A typical overall frequency response curve for the Zenith color amplifier section is shown in Figure 7.6. This same response curve, with only a minor variation, is used in practically every recent color TV receiver. Note that the 3.58 mHz color sub-carrier is not in the exact center of

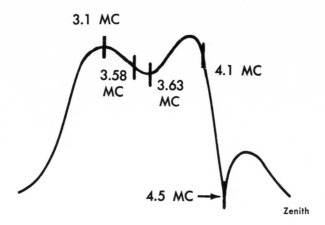

Figure 7.6—OVERALL COLOR AMPLIFIER RESPONSE

the response curve and that the high frequency end, towards the 4.5 mHz intercarrier sound, is very steep. It is very important that the intercarrier IF be attenuated sufficiently to avoid the beats which could otherwise result between it and the color sub-carrier. The general rule is that the 4.5 mHz must be at least 55 dB below the peak of the color sub-carrier response curve, that the valley between peaks shall not be more than 10% and that the peaks should be separated by approximately 1 mHz. In most color TV receivers the response curve of the color sub-carrier or chroma amplifier is quite flat and symmetrical.

The detailed operation of the demodulators is the subject of Chapter 9 but the principles of operation can be described briefly as follows. In the demodulator the color sub-carrier is compared against the constant phase and fixed amplitude of the color synch frequency, 3.58 mHz. At any one instant the output of the demodulator will depend on the difference between the phase and amplitude of the color sub-carrier and the color synch signal. The color sub-carrier phase and amplitude vary with the hue and saturation of the color component which makes up the particular picture element. The video signal, resulting from the phase and amplitude comparison in the demodulator, is of a much lower frequency content than the sub-carrier. As we have seen in the above paragraph, the bandpass response of the color sub-carrier amplifier is only ± 0.5 mHz. As a result, the video signal coming from the modulator will have a frequency content from approximately 60 Hz to 0.5 mHz. To remove the 3.58 mHz color sub-carrier and reference signal components, a 3.58 mHz trap is used to make sure that the 3.58 mHz component is kept away from the color picture tube itself.

Chapter 9 also describes the theory and detailed circuitry which is used to matrix or add the various vector signal components from the two demodulators to obtain three color difference signals. In effect, however, it is apparent from the vector diagram of Figure 3.9 that any two vectors, such as the X and the Z or the R-Y and B-Y vectors, can be used to derive the third vector. Simple vector addition as illustrated in Fig. 3.3 is used in all color TV receivers to obtain the third color. Vector addition appears simple in two-dimensional geometry and the electrical circuitry itself is also easy to understand but requires careful considerations of phase, polarity, and amplitudes.

As shown in Fig. 7.2 and Fig. 7.3, the horizontal and vertical deflection circuits of most color TV sets are not materially different from their monochrome counterparts. For this reason we will omit them in this discussion and turn to the color synchronizing circuits.

The first stage in the color synch system must remove the 8-cycle burst of 3.58 mHz color reference signal from the blanking pedestal of each horizontal synchronizing pulse. This is accomplished by providing a keyed amplifier, whether transistor or vacuum tube, which receives the composite video signal on its input and a keying pulse from the horizontal oscillator. In transistor circuits, the video signal is usually applied on the base and the keying pulse is applied to either the emitter or collector. In vacuum tubes, the video signal goes to the grid and the keying pulse is applied to the cathode, the plate, or, in some multi-grid tubes, to the suppressor grid. In either circuit only the amplified 3.58 mHz reference carrier burst appears at the output. This reference burst is used to control the frequency and phase of the color synchronizing signals generated in the color synch section.

In some receivers the local oscillator is stabilized by a quartz crystal at 3.58 mHz and the reactance stage controls the frequency only over a limited range. Other color AFC circuits do not make use of a crystal for the main tuning element and still other systems employ a crystal ringing circuit without local oscillator and with AFC. All signals in this section are 3.58 mHz and have a sinewave shape.

All color receivers have an automatic circuit, usually called the color killer, which senses whether a color or a monochrome transmission is being received. When a monochrome transmission is received, it cuts off the color amplifier portions, and, in most receivers, the color synchronizing portions. If this circuit does not operate properly and the color sub-carrier amplifier and color synchronizing circuits remain active, they can generate noise which will appear as colored sparks flying across the screen, superimposed over the monochrome picture.

The color killer circuit determines whether a monochrome or color transmission is being received by sensing the output of the color burst amplifier. On all color transmissions an 8-cycle burst of 3.58 mHz color reference signal is transmitted on each horizontal synchronizing pulse and this signal is then available at the output of the burst amplifier. On monochrome transmission the output of the burst amplifier is zero. A variety of different color killer circuits are discussed in Chapter 10 and more detailed descriptions of their operation are presented.

A number of other circuits of a special nature are found in color TV receivers which are not directly associated with color transmission but which are necessary to produce correct black and white pictures as well. Among these circuits are the convergence section and the pin cushion correction section, both of which will be discussed again in connection with the deflection circuits in Chapter 11.

Monochrome Reception

When black and white telecasts are received, the overall circuit operation is much simpler than during color reception. The RF and IF stages operate just like in any TV receiver and the detected video and audio signals are fed to their respective amplifiers. On monochrome reception the IF response curve, trap alignment, and sound IF rejection are not as critical and misalignment may not be apparent. The video signal goes through the first video amplifier; the vertical and horizontal synchronizing signals are separated and supplied to their respective sweep sections. Since there is no color synchronizing burst present, the burst amplifier will not furnish any signal to the color phase detector and this will cause the color killer to cut off the bandpass amplifier. No signal passes through the demodulator, amplifier and phase splitter stages. The color reference oscillator is still operating in some sets, but in some other receivers provisions are made to shut the oscillator off during monochrome transmission.

The video signal goes from the first video amplifier to the second video amplifier after passing through the delay line. This delay of about

1 microsecond does not affect the transmission in any appreciable manner. Proportionally equal amounts of video signal are applied to the three electron guns through the red, green, and blue matrix stages. When they are properly balanced, the result will be a white, gray, black picture without any colored portions in it.

For monochrome operation the convergence and focus correction circuits are as important as for color reception. When purity, convergence, or focus are misadjusted, it will be impossible to obtain a clean black and white picture and some colors will always be present.

Color Reception

As pointed out above, the entire decoder section is deactivated during monochrome reception. This is accomplished automatically through the color killer circuit. During color operation, all parts of the receiver are working and each individual adjustment or misadjustment will contribute something to the picture on the screen. When properly aligned and adjusted, a color set should require no manual readjustment when switching from a black and white to a color telecast or back to monochrome.

Sometimes it may look as if one of the color controls needs to be reset, but if the entire receiver is working correctly, none of these controls should need attention. Many people prefer to view a monochrome telecast with a bluish rather than a true white screen, and then, when color pictures are viewed, too much blue may appear in the color image. Another frequent trouble source is the tendency to adjust for too much brightness during a color program, resulting in HV overload for monochrome. In general it holds true that whatever misadjustments take place in monochrome reception, they will appear much more objectionable when a color telecast is received. A good example of this is the case in which slight misadjustment of the fine tuning control occurs. This may go unnoticed on monochrome operation, but may show up as a fine beat interference or as poor color quality on a color telecast.

In servicing color TV receivers it is important to realize that some of the B-plus voltages will be different for monochrome and for color operation. The RCA service manuals specify that all voltages are measured with a black and white test pattern which is applied at the antenna terminals and which is 500 microvolts strong. This last specification is important because it gives an indication as to the AGC conditions and therefore to the current drawn by the IF stages. Many manufacturers are now stating operating voltages for both monochrome and color reception since color telecasts are now available everywhere.

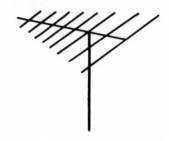

ANTENNA, TUNER
AND IF SECTION

The basic functions of the three "front end" portions of the color receiver are the same as for the monochrome TV or any other superheterodyne type of receiver. The antenna intercepts the RF signal, the tuner selects the desired channel and heterodynes the signal down to a fixed IF. In the IF section the signal is amplified sufficiently before being detected. Gain and bandpass characteristics for black and white receivers have been pretty well standardized and the circuits employed will be familiar to the reader. For color reception the requirements are somewhat more stringent and more complex circuitry is needed.

Bandpass Requirements

In color TV these three "front end" sections must operate very close to the desired response characteristic because of the presence of the color sub-carrier and the fact that both amplitude and phase modulation are employed. A change in frequency response or a change in gain may result in a different color signal and therefore in wrong colors on the screen. The color decoder section cannot distinguish between phase modulation originating at the transmitter and phase shift due to receiver circuitry. An example of such a color error is shown in the illustration of Figure 8.1. The correct overall response curve of the

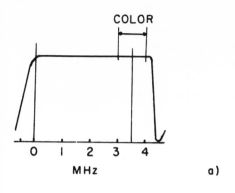

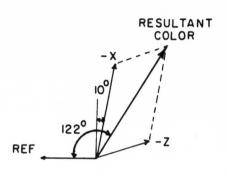

a)

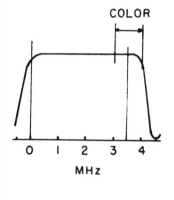

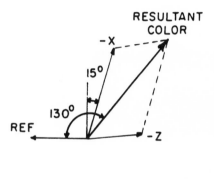

b)

Figure 8.1—a, CORRECT OVERALL RESPONSE AND DECODED COLORS; b, INCORRECT
OVERALL RESPONSE

antenna, RF and IF section is shown in Figure 8.1a together with the
decoded colors in vector form. Below, in Figure 8.1b, drop in amplitude
at the edge of the color sub-carrier side band results in phase shift. Since
the color visible on the screen is the added value, it is apparent that
the response curve of Figure 8.1b will produce the wrong color. Theoretically, the slope of the response curve will control two aspects of
the amplified signal, the amplitude and the phase delay. The phase
delay is the time difference between the signal at the input to a
particular stage and at the output. Although electrons travel at the
speed of light, the difference between two signals having different
delay time is usually stated in electrical degrees. Returning to fundamental electrical concepts, we know that in an inductive circuit the

voltage leads the current by a given angle. For a pure inductance the "phase angle" between voltage and current is 90 degrees, and when some resistance is present the angle will be less than that. Another basic concept is that the signal at the plate of an amplifier tube or the collector of a transistor is 180 degrees out of phase with the signal at the grid or base. In a tuned amplifier the exact 180 degree phase difference will hold true only for the resonant frequency, and signals occurring at the slope of the response curve will be shifted a somewhat different amount. If a 4 mHz passband is needed for correct color reproduction this means that the phase shift of all signals within this passband should be exactly the same. The phase delay characteristic of the amplifier system would then be flat over the 4 mHz passband.

This phase delay characteristic is a criterion just as the bandpass or amplitude versus frequency characteristic and a typical delay versus frequency curve is shown in Figure 8.2. Note that the horizontal axis

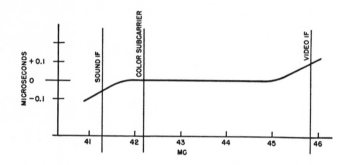

Figure 8.2—TIME DELAY VERSUS FREQUENCY CURVE

is calibrated in terms of frequency, the IF band in this example, while the vertical axis represents phase shift in microseconds. Just as the overall IF bandpass response curve is the sum of the response curves of the individual IF stages, so is the phase delay characteristic the result of all the individual phase delay characteristics. Figure 8.3 shows the phase delay in degrees for four different frequency signals, representing the Y IF carrier, the IF signals corresponding to the I and Q, and the 3.58 mHz color sub-carrier.

The frequencies shown here for the I and Q signals do not represent these signals in themselves but designate the maximum bandwidth ascribed to them   at the transmitter. With the limited bandwidth, ± 0.5 mHz, used in the demodulators of practically all color TV receivers, the phase shift effects will be much less for those signals closer to the 43 mHz point. In the case of the demodulated color signals which go below the 42.1 mHz IF color sub-carrier frequency phase delays will have a pronounced effect.

In Figure 8.3a the phase delay of the first IF stage is shown for these specific frequencies. Under b and c the delay characteristics of the second and third IF stage are shown and the dotted line illustrates the overall phase delay of the total of three IF stages. It becomes apparent that a shift in this phase delay at any one stage might affect only one of the four signals used in this illustration, but since the color of the viewing screen is due to correct decoding and matrixing of three signals the wrong color may appear.

Figure 8.3—PHASE DELAY FOR 4 SIGNALS

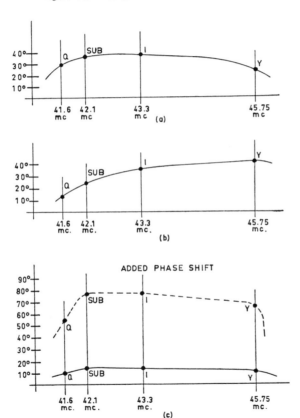

The bandpass and phase delay characteristics of a circuit are not linearly related to each other and the only safe criterion of judging phase delay by bandpass response is to say that along the flat portion of the response curve the phase delay is uniform. For this reason great emphasis is placed in color TV receiver alignment on obtaining a really flat top, wide band frequency response. It is possible, however, by wrong design to build an IF section which has an overall flat response curve but which has non-uniform phase delay over the passband. This can be due to certain combinations of Q, gain and phase shift in the different IF stages. Ordinarily such a defect will not occur in a commercial color TV receiver if each IF stage is tuned to its specified frequency. Servicemen often align a monochrome TV IF section without regard to the manufacturer's data, simply to obtain a decent response curve on the oscilloscope. In color TV alignment this practice cannot be followed without running into real trouble.

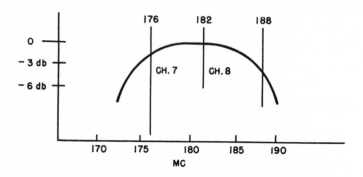

Figure 8.4—FREQUENCY RESPONSE OF NARROW BEAM ANTENNA

Although the IF section contributes the greatest amount of amplification and therefore is the major factor in determining phase and bandpass characteristics, the antenna and tuner sections also have considerable effect. Most TV antennas are broad enough in frequency response to pass any 6 mHz channel without appreciable phase delay variations. Ordinarily antennas used for all VHF channels have a fairly flat response over the entire band. Yagi and other directional antennas, however, often have a narrow RF response curve and can cause difficulties in color reception. Figure 8.4 shows a typical frequency response curve of a stacked, directional fringe area type antenna. Assume that originally this antenna was designed for channels 7 and 8, a practice followed earlier by some manufacturers to make it possible to sell one model in many different areas. For monochrome the response of the antenna on channel 8 was adequate since the gain was only 3 db down at the sound RF carrier. For a color signal, this situation may cause some difficulty because the edge of the sub-carrier side band reaches into the 3 db drop off region of the antenna response curve. As a result, the phase delay and amplitude relationship between color signals at that point and the color sub-carrier itself will no longer be correct and poor color fidelity may result.

In the UHF channels the antenna and tuner frequencies are so much higher that the relative bandwidth and therefore the flatness of the phase response is not usually a problem. At 500 mHz for example, a bandwidth of 6 mHz will be obtained even from a highly directional, sharply tuned Yagi antenna.

In general, antenna bandpass characteristics only become troublesome in fringe area installations or in instances where considerable mismatch exists between the antenna, transmission line and receiver. Impedance mismatch, causing standing waves, is a frequency sensitive phenomenon and therefore troublesome in color TV.

This is the main reason why in installing a color TV receiver the technician should be sure to use properly matching antenna and transmission line, and employ matching transformers, tuning stubs, and similar devices to remove standing wave defects. When a color TV set is installed in place of a monochrome receiver, be sure that the monochrome installation does not contain mismatch or a detuned antenna.

Another frequent reception difficulty is the appearance of "ghosts." Disturbing in monochrome, multiple images become intolerable in a color picture. Every effort must therefore be made to eliminate multiple path signals by using directive, narrow beam antennas and proper impedance match at all frequencies.

The problems of ghosts in color TV reception are quite severe and often appear insurmountable. Chapter 12 contains a special section, with references to actual color photographs, which discusses the problems of color ghosts and suggests a number of methods for eliminating or reducing this type of interference.

The customer should understand that a color installation is more expensive, because closer tolerances must be maintained to get satisfactory color pictures. The RF amplifier and mixer in the color receiver tuner section are usually the same as for monochrome both in circuitry and in alignment. Manufacturers tighten the tolerances on bandwidth and peak-to-valley ratio in their alignment procedure, but the desired response curve is the same. A typical tuner RF response curve is shown in Figure 8.5. Since the gain is relatively small in the RF stage as compared to the overall amplification, the phase delay due to that stage is fairly uniform and does not require too much attention.

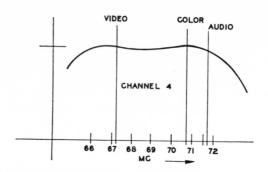

Figure 8.5—TYPICAL RF TUNER RESPONSE CURVE

Sound Carrier Attenuation

The bandwidth and phase delay requirements of the IF section are only two of the unique problems encountered in color TV receivers. Rejection of the sound IF carrier down to the proper level presents another problem to the designer and correct alignment of the sound traps is an important operation for the serviceman. In intercarrier monochrome TV sets the sound carrier is usually about 30 to 40 db down from the flat portion of the response curve as shown in Figure 8.6a. This is necessary to prevent the sound signal from appearing in the picture, a phenomenon frequently observed when the fine tuning control is misadjusted. The 30 or 40 db sound rejection is obtained by using one or two absorption or series type traps somewhere in the IF section. These traps are tuned to the sound IF and have sufficient Q to do the job. In color TV the IF response should be flat to within 400 kHz of the sound IF carrier and this means that, even for 40 db sound rejection, a trap having a narrow response must be used. Actually the amount of sound rejection for color TV IF sections is in the order of 60 db, making the problem even more complex.

To understand the need for so much sound IF rejection, consider the fact that at the detector all signals present are detected and produce beat signals. The beat between the video and the sound IF carrier is 4.5 mHz and this is usually too high in frequency to cause severe interference in the picture. Between the sound carrier and the color subcarrier the beat signal is only 920 kHz, which is quite visible and objectionable in the picture. Such a beat signal will be amplified along with

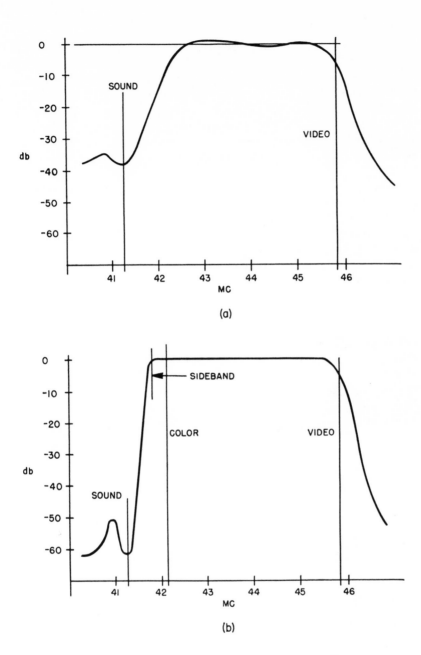

Figure 8.6—a/ MONOCHROME IF SOUND REJECTION CURVE; b/COLOR IF
SOUND REJECTION CURVE

the desired video band. No special trap can be used to eliminate it as is sometimes done for the 4.5 mHz beat. Beats between the sound carrier and the various sidebands of the color sub-carrier add further to the need for at least 60 db of attenuation for the sound IF. When such great attenuation is obtained, the 4.5 mHz intercarrier IF will also be much weaker than in the monochrome counterpart and therefore many of the current color receivers use separate sound detectors, tuned to the sound IF carrier and followed by an extra stage of 4.5 mHz amplification.

The response curve of Figure 8.6b shows a typical, correct IF response curve with sufficient sound rejection. Note that the sound carrier frequency, 41.25 mHz, falls into a really deep and narrow slot. This narrow rejection band cannot be achieved simply by adding more sound traps to a monochrome type IF circuit. Although additional traps would increase the amount of rejection they would also widen the rejection band since their effect is additive at all points of the response curve. If it were possible to use traps having a much higher Q, the sharp rejection notch could be achieved, but it is not practical to use such coils in the crowded TV receiver chassis.

To get sufficient sound carrier rejection most color receivers use some type of multiple section filter, usually the "Bridge T" type shown in the circuit of Figure 8.7. In this filter C1-L1 are tuned to the IF frequency while L2 is tuned to the rejected frequency. A more complex form of trap is shown in Figure 8.8. This unit actually requires three adjustments, two for the sound IF carrier and one for the adjacent sound IF carrier. The combination L1, C1, and C4 are series resonant

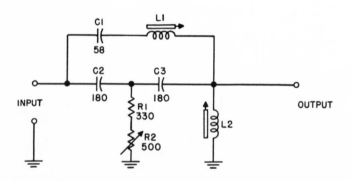

Figure 8.7—BRIDGE-T TYPE FILTER

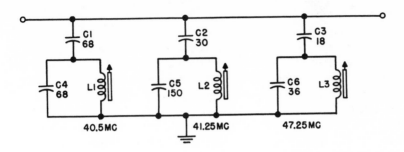

Figure 8.8—COMBINATION SOUND AND ADJACENT CHANNEL TRAP

and present a short circuit for 40.5 mHz. Three such networks are used in a typical GE receiver. Some manufacturers prefer to use single bifilar transformers with the trap wound right along with it. Such a circuit is shown in Figure 8.9 and depends largely on the proper design of the coils and the location of the traps for good rejection. Various similar sound rejection circuits are used, and in each instance the alignment of the sound IF trap is of great importance and must be done accurately and carefully. Since the frequency tolerances are so small, it is essential

Figure 8.9—BIFILAR SOUND TRAP

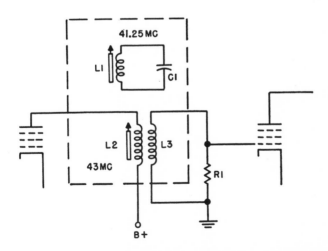

that the signal generator used for alignment be quite accurate, preferably crystal controlled. Most of the current TV alignment generators used for monochrome work have crystal check points to allow sufficient frequency accuracy for IF trap alignment.

Some of the IF sound traps, as shown in Figure 8.8, are stagger tuned and their alignment must be done carefully to make sure that the right coil is tuned to the right frequency. It is also important to remember that usually the tuning of one coil has some effect on the nearby circuits. After each major adjustment all coils in one trap assembly should therefore be touched up. Although emphasis is placed here on the alignment of the sound rejection circuits, the adjacent channel traps are also important. In general these traps are tuned to the adjacent channel sound and video IF just as in monochrome TV receivers.

Oscillator Drift

Having made all the required adjustments carefully and correctly, the serviceman is still not assured of good color fidelity pictures. The remaining step is to make sure that the received signal actually occupies the available IF passband. In other words, the alignment and fine tuning adjustment of the local oscillator determines whether the respective carriers will really fall into their appointed places in the IF band. By correct adjustment of the local oscillator frequency the received signal will be heterodyned down to the desired IF. The problem of keeping the oscillator frequency stable after warm up and in spite of tube aging has been the subject of much study. To understand the actual limits within which oscillator drift will not become troublesome, refer back to the overall IF curve of Figure 8.6b. Note that the color sub-carrier is at 42.17 mHz and the end of the flat portion of the curve is at 41.65 mHz, leaving 520 kHz of flat portion for the 500 kHz bandwidth of the color sub-carrier side bands. In order to avoid beats and loss of color fidelity, the 500 kHz space should remain completely on the flat portion of the response curve, allowing the local oscillator to drift 20 kHz from the optimum point before the picture deteriorates.

Permissible oscillator drift in the other direction is also limited. Although only the brightness signal and the synch pulse amplitude will suffer at the high end of the IF band, the sound IF carrier will no longer fall into the narrow notch and will be amplified sufficiently to cause a beat with the color signals. When the local oscillator has

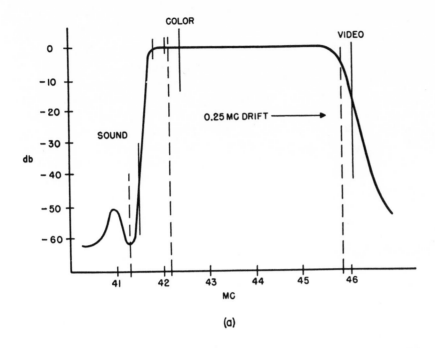

(a)

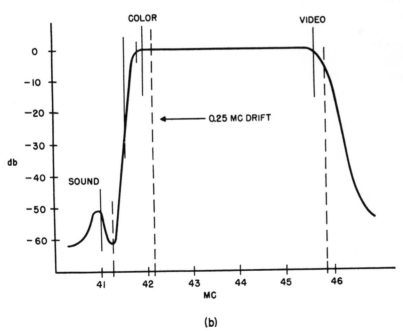

(b)

Figure 8.10—a/IF RESPONSE WITH HIGH OSCILLATOR FREQUENCY b/
IF RESPONSE WITH LOW OSCILLATOR FREQUENCY

97

drifted too high in frequency the sound and video IF carriers also will be higher and shift the actual IF response as shown in Figure 8.10a. When the local oscillator frequency is too low, the response curve of Figure 8.10b is obtained. Although either of these mistuning defects can occur, the most frequent case is the one in which the oscillator frequency drops after warm-up or tube aging and then color beats and loss of color fidelity are the trouble rather than sound interference. When the oscillator drifts sufficiently to bring the sound carrier substantially above the rejection notch, then its beating will add to the other picture defects.

In order to avoid excessive oscillator drift several different designs have been perfected. One stabilizing circuit makes use of the sound IF detector to combine it with an AFC system which controls the local oscillator frequency. Another system employs a very stable oscillator circuit, and, by adding temperature compensating capacitors, the required stability can be obtained. Crystal controlled RF oscillators have also been proposed, but their expense and complexity hardly warrant their use. Finally it has been pointed out that as long as the fine tuning control has sufficient range to compensate for oscillator drift, the viewer can clear up the trouble himself after warm-up. In most VHF tuners the fine tuning control has a range of about 0.5 mHz, which should compensate for oscillator drift. If, however, the oscillator is aligned barely to reach the correct frequency at a cold start, the fine tuning control may not be sufficient to compensate for warm-up drift. In monochrome receivers the small loss of fine detail due to a shift of 50 or 100 kHz was not usually noticeable, but in color reception the range of the fine tuning control and the oscillator adjustment for a cold start and after the warm-up period are quite important.

So far we have considered only the oscillator stability in VHF tuners. Actually the problem on UHF is much more severe. At these higher frequencies the percentage drift of the oscillator might be the same as in VHF, but because of the higher operating frequency the actual drift in mHz is often excessive.

The almost universal use of a transistor as local oscillator for the UHF tuner has limited the temperature problem somewhat. Temperature compensating capacitors have further reduced it. It still remains true, however, that many UHF tuners require frequent fine tuning touch-ups on color reception. The increasing use of transistors in VHF tuners as well as in IF stages promises to reduce the tuner problem and the more widespread use of automatic frequency control circuits in tuners also helps.

Typical IF Circuits

Just as in monochrome TV receivers and in any kind of superheterodyne receivers, the IF section of color TV sets must perform two basic functions. It must amplify the very weak intermediate frequency signal received from the tuner and it must limit the bandpass of this signal. The need for the accurate bandpass response has already been discussed and Figure 8.6b illustrates the desired IF response for color TV receivers. The circuits used to provide this amplification and bandpass response range from the old fashioned vacuum tube amplifiers to transistors and, the very latest, integrated circuit IF amplifiers. Regardless of which circuit is used, the overall amplification must range from 60 to 80 db and the bandpass response must closely approach that shown in Figure 8.6b.

A transistorized IF section which is part of the Philco line of color TV receivers is shown in Figure 8.11. At the input from the tuner there is a special trap assembly consisting of L11, L12, and L13 and their associated capacitors. L11 and L14 are series traps while L12 is a transformer coupled to L13 which is series tuned to 47.25 mHz, the adjacent channel sound IF. L12 is tuned to 38.75 mHz, the adjacent channel video IF. This trap is a little broader and will also reduce the 41.25 mHz sound intercarrier. A special trap for the 41.25 mHz intercarrier sound is provided by L6, which is in series with the 1N60 video detector. In the transistor circuit of Figure 8.11 the sound rejection is further controlled by coil L2 and a 750 ohm potentiometer connected across it. Depending on the setting of this potentiometer the self-resonant coil L2 will tend to reduce the 41.25 mHz intercarrier sound IF.

The coupling transformers between the first and second, and the second and the third transistor amplifiers are somewhat different from those used in vacuum tubes but they are essentially the same, including the trap, as those used in all other transistor IF amplifiers. Nor is the AGC circuit unique to color receivers. Practically all transistor IF amplifiers use a reflex type or bootstrap of AGC system as illustrated in Figure 8.11. The main AGC voltage from the second AGC amplifier Q3 is applied to the base of the second IF transistor Q5. A portion of the emitter current of that stage is fed back to the base of the first IF stage so that any change in bias in the second IF stage will result in a change in emitter current and therefore a change of bias of the first IF stage. This arrangement provides a small amount of AGC amplification and at the same time provides more linear control of the gain of the first two stages.

The output of the third amplifier is supplied to the video detector and to a separate sound IF detector. The sound detector obtains its signal directly from the collector from the third IF stage, with diode D4 providing the detection and the T-filter, consisting of L1, L5, and C6, removing the IF components.

A typical IF amplifier using a single integrated circuit chip is illustrated in Figure 8.12. This circuit contains three separate amplifier stages for the video IF, a video detector, and a video preamplifier. It also contains a sound IF amplifier, a detector, and a preamplifier for the 4.58 mHz intercarrier sound. In addition, the same chip contains AGC amplifiers and a zener voltage regulator circuit.

The IF signal from the tuner passes through a series resonant trap, tuned by L1, and a bandpass-shaping, double-tuned transformer, T1, from which the IF signal goes to pin 6 on the IC. The output of the AGC circuit is obtained from pin 4, and this is applied, through a series of resistors, to the input signal at pin 6. Note that the gain adjustment controls the AGC voltage through the 100K resistor.

Between the first IF stage and the second IF stage, the signal goes to the tuned circuits consisting of T2 and L3. Pin 12 carries the input to the separate sound IF section which has the 4.5 mHz sound output at pin 2. The output of the first IF stage is at pin 9, and the input to the second video IF stage is at pin 13. Between these two points there is a tuned circuit consisting of L2 and a 3.9 pf capacitor and the coupling transformer, T2, which provides the sound take-off to the separate sound IF amplifier at pin 12. The resonant circuit, consisting of L3 and the 120 pf capacitor, is a trap to reject the sound IF from the second video IF stage. For keyed AGC, the horizontal keying pulse is applied to pin 3 through a resistance network, not shown in Figure 8.12.

The single transistor shown in Figure 8.12 acts as power supply isolating and voltage regulating circuit and assures that the IF IC will be immune from voltage variations in the rest of the color TV receiver. Note that, on pin 14, an output is available for an automatic fine tuning (AFT) circuit which is used to control the VHF and UHF tuners. At pin 7 a delayed AGC voltage is available which goes to the RF amplifier in the VHF tuner. The video output signal at pin 19 is the detected signal, passed through one stage of amplification, and is usually in the order of 7 volts peak amplitude.

In Figure 8.12, a typical IC IF amplifier circuit using the R.C.A. type CA 3068 color TV IF system has been described. This circuit provides a gain of 75 db and has an inherent bandwidth capability

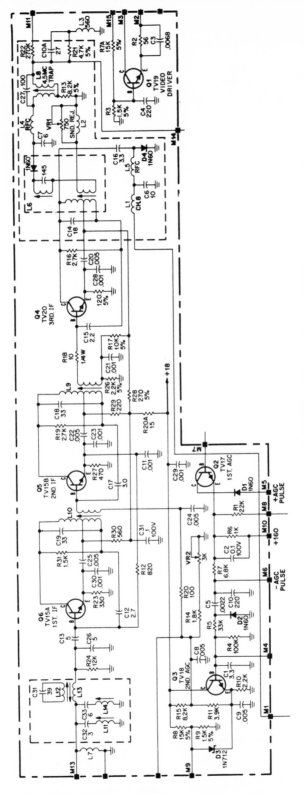

Figure 8.11—IF SECTION OF PHILCO P-LINE COLOR SETS

101

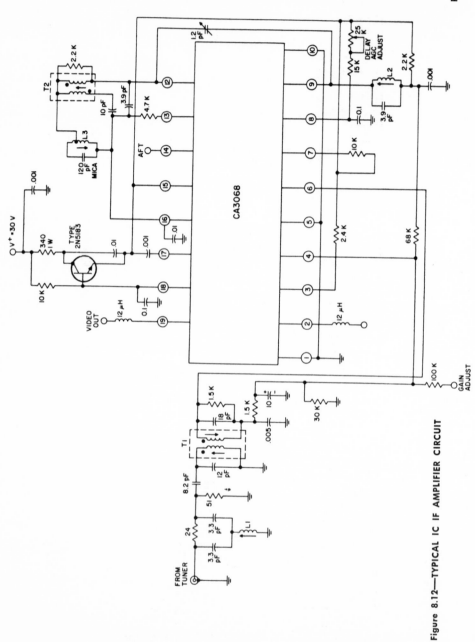

R.C.A.

Figure 8.12—TYPICAL IC IF AMPLIFIER CIRCUIT

102

between 10 and 70 mHz. It is all contained on a single, 20-pin, in-line plastic package and is typical of the IC sections used in many advanced color TV receivers.

9

THE COLOR DECODER

Decoder Functions

This chapter describes the circuits which remove the color information from the color sub-carrier and supply it to the color picture tube for display. Chapters 3 and 4 have explained the method of encoding the color information on the color sub-carrier and the steps necessary to decode this information. In order to obtain the desired red, green and blue video signals a number of separate operations must be performed. First, the color sub-carrier and its side bands must be separated from the composite video signal and amplified to the desired voltage levels. To decode the color information carried on the sub-carrier the 3.58 mHz reference signal must be compared with the phase and amplitude of the color sub-carrier. This process of phase and amplitude comparison is new and unique to color TV and does not have any comparable function in other consumer equipment.

A variety of different circuits are used to accomplish this and the main portion of this chapter will be devoted to an explanation of these complex but very important circuits. Once the phase comparison is accomplished, the results will be two video signals which represent some kind of color difference information. "Color difference" means that the brightness component is not contained in this video signal. The third step in the decoding process is called matrixing and consists of performing vector addition and subtraction, by means of electronic circuitry, so that the decoded difference signals are converted into the red, green and blue color difference signals. These color difference signals are then added to the brightness signal in the color picture tube itself, where they generate the three primary colors in the form of electron beams.

The different steps required for decoding are performed in three basic circuits: the bandpass amplifier, the demodulator and the matrix. In some recent color TV receivers the demodulator and matrix network appear as one circuit section and it may be difficult to find where one ends and the other begins. In every case, however, the functions of decoding and matrixing, can be clearly defined.

Figure 9.1 shows the block diagram of a typical chroma or color demodulator section. In this particular section, which is characteristic of a very large number of today's color TV receivers, the X and Z vectors shown in Figure 3.9 and described in Chapter 3, are used as references.

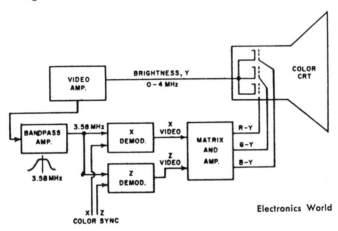

Electronics World

Figure 9.1—BLOCK DIAGRAM OF TYPICAL CHROMA DEMODULATOR

Beginning at the video amplifier, the signals are separated. The brightness or Y signal, which contains all of the detailed picture information and which is the only one that is available on monochrome transmission, goes directly from the video amplifier to all three kinescope cathodes. The color sub-carrier is removed from the composite video signal by the bandpass response of the bandpass amplifier, sometimes also called the chroma amplifier or the color IF. This stage then provides the 3.58 mHz sub-carrier signal to the X and to the Z demodulator stages. In each of these stages the color sub-carrier is compared, in amplitude and phase, with the fixed phase and amplitude X and Z color synch signal. The output of the demodulator is the X and Z video signal which is then supplied to the matrix and amplifying section. By manipulating the X and Z video signals in a vector addition and subtraction process the matrix section generates the red, green, and blue color difference signals. These signals are then supplied to the control grids of their respective electron guns. With the brightness signal supplied to the cathode and the color difference signal applied to the grid, addition takes place in the electron guns directly.

In some recent color TV receivers the X and Z demodulators are replaced by R-Y and B-Y demodulators which then produce the R-Y and B-Y signal directly. By vector addition and subtraction the G-Y signal is generated from the other two. Occasionally, the G-Y signal is decoded directly, together with the R-Y and B-Y, by identical decoding circuits. In some color TV receivers, particularly those using a single IC for the demodulator, the brightness signal is added to the three color difference signals before they reach the color picture tube control grids. The major differences between the X, Z or R-Y, B-Y decoding lie in the circuitry that is used for the demodulator and matrix section, and in the phase angle of the color synch signals which determine whether the X and Z vectors or the R-Y and B-Y vectors are decoded.

In the table of Figure 9.2 the vector relationships between the red, green, and blue color difference signals, I and Q, and X and Z are shown in simple algebraic form. The technician need not memorize the exact values of the various signals but the operation of the demodulators and the matrixing circuits is based on these amplitude and polarity relationships. In particular the polarity is important because it determines whether a signal must pass through an additional stage of amplification

Figure 9.2—TABLE OF COLOR VECTOR RELATIONS

Red = Y + 0.95 I + 0.62Q

Green = Y - 0.27 I - 0.64Q

Blue = Y - 1.1 I + 1.7Q

$R - Y = -X \cos 10.9° = -0.982 X$

$R - Y = -Z \cos 73° = 0.292 Z$

$B - Y = -Z \cos 17° = 0.956 Z$

$B - Y = -X \cos 79.1° = -0.189 X$

$G - Y = X \cos 23.1° = 0.919 X$

$G - Y = Z \cos 30° = 0.777 Z$

Vector addition:

$1(R - Y) = 0.21 Z - 1.16 X$

$1(B - Y) = 1.11 Z - 0.34 X$

$1(G - Y) = 0.41 Z + 0.71 X$

$1(G - Y) = -0.54 (B - Y) - 0.84 (R - Y)$

or whether it should be applied at the grid or at the cathode, the base or the emitter, of a circuit.

The Chroma-Bandpass Amplifier

As indicated in the block diagram of Figure 9.1 the entire video signal from the video amplifier goes to the bandpass amplifier and from there only the sub-carrier goes to the two demodulators. The functions of the bandpass amplifier can be understood by referring to the simplified diagram of Figure 9.3. The input to the amplifier circuit is the entire video spectrum, from 60 Hz to approximately 4 mHz. Located within this video band is the color sub-carrier at 3.58 mHz, and its sidebands. It must be remembered that the color sub-carrier and its sideband are transmitted at much less amplitude than the brightness or Y. This is one of the reasons why a certain amount of gain is necessary. Another reason is that the color sub-carrier must be separated from the brightness signal at a point when both are at a relatively low level since otherwise intermodulation distortion can occur.

Before discussing typical circuits, the main function of the bandpass amplifier should be discussed because, regardless of whether the particular TV receiver uses vacuum tubes, transistors or integrated circuits, the functions of the bandpass amplifier remain the same. As illustrated in Figure 9.3 there must be at least two tuned circuits, L1 and C1, and L2 and C2, to provide the bandpass required. At the input to the amplifier the tuned circuit must eliminate all of the lower frequency components as well as the higher frequency components. It is especially important that the 4.5 mHz intercarrier sound IF is attenuated considerably since it can produce very annoying patterns. In effect, the bandpass amplifier is a tuned amplifier, which has its gain controlled by the color killer bias.

Figure 9.3—BANDPASS AMPLIFIER FUNCTION

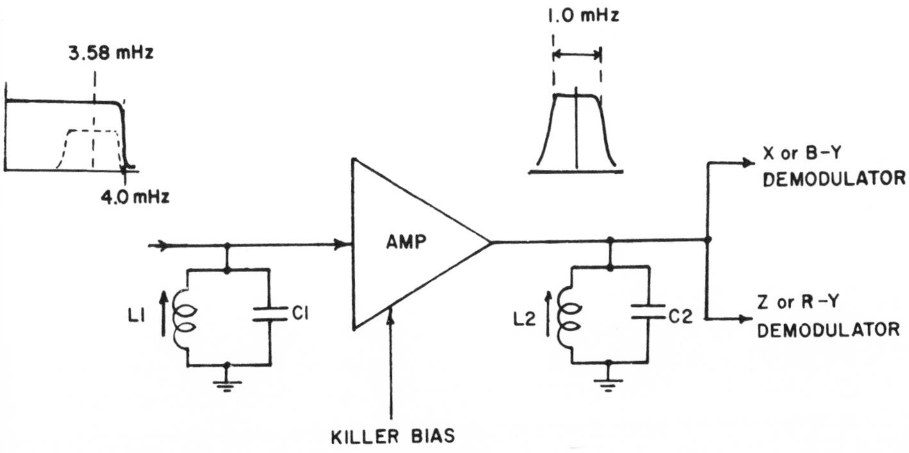

When a monochrome transmission is received, the color killer bias will turn the bandpass amplifier effectively off. In some of the most recent color TV receiver models a second control bias, not shown in Figure 9.3, goes to the amplifier. This bias provides automatic gain control for the color or chroma signal itself, similar to the AGC system of the video IF section. The output of the bandpass amplifier invariably goes to the demodulators. Depending on the circuit, this demodulator may either produce the X and Z or the R-Y and B-Y vectors directly.

To illustrate how the functions described in Figure 9.3 are performed in a typical color TV receiver, the chroma circuit of the Sylvania model D16 series is shown in Figure 9.4. A total of three stages of amplifica-

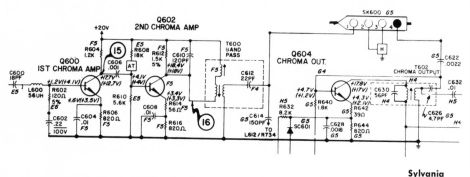

Sylvania

Figure 9.4—TRANSISTOR CHROMA AMPLIFIERS

tion are used here. The first chroma amplifier stage, Q600, receives the weak signal at its base, through the peaking coil, L600. The second chroma amplifier, Q602, provides additional gain and contains the bandpass transformer, T600, which couples the amplified 3.58 mHz chroma signal via connector SK 600 to the color gain control, not shown in Figure 9.4. Pin 4 of the same connector returns the signal to the base of Q604, the output stage. The emitter bias of Q604 is controlled by color killer bias, applied through the 8.2K resistor R632. When monochrome transmissions are received, this color killer bias cuts off Q604. The diode SC601 is in the circuit to assure the proper DC levels.

The typical problems of adjusting the chroma bandpass response are illustrated in Figure 9.5. Although these response curves originate in a vacuum tube circuit, the effect of the tuning adjustments shown in Figure 9.4 are essentially the same. The notations concerning the tolerances, such as "SAG 20% ± 10%," are indications of the alignment accuracy required. In general, the alignment is not too critical, and, once it is properly set, rarely requires adjustment.

Figure 9.6 shows a somewhat different color IF section, or bandpass amplifier, as used in a Motorola color TV model. From the first video amplifier a color signal is applied to the control grid of one half of the

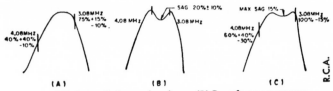

**(A) Over-all chroma bandpass. (B) Transformer response.
(C) Grid coil response.**

Figure 9.5—CHROMA BANDPASS RESPONSES

6DX8 triode. This tube acts as cathode follower for the color synch and for the 3.58 mHz color sub-carrier. The sub-carrier signal is brought to the color intensity control through a frequency compensating network, L900 and L901, and a coaxial cable. This control, frequently called chroma gain, is located at the front of the color receiver. L902 helps to maintain the frequency response from this color intensity control to the second stage of the bandpass amplifier.

Some self-bias is provided in the cathode, but the gain of the tube is controlled largely through the grid voltage, the killer bias applied across R903. The color killer circuit itself will be described in Chapter 10 but at this point it is only important to understand that, with no current drawn through the color killer triode, no voltage will be developed across R903 and the grid bias of the bandpass amplifier pentode will be only that due to the self-bias cathode circuit. The output of the bandpass amplifier is transformer T900 which contains a number of features

Figure 9.6—DETAILED CHROMA AMPLIFIER CIRCUIT

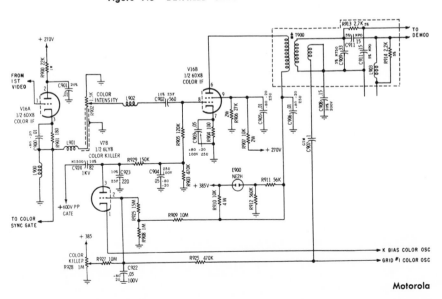

not found in many other sets. One of these is an extra winding which couples a portion of the signal to the control grid of the color oscillator. Still other features are the RC networks in series and shunt with the T900 secondary. These networks affect the phase of the two signals going to the demodulators. These features will be discussed in Chapter 10.

The circuit of Figure 9.6 contains only one sharply resonant circuit, T900, and this is tuned by a single slug. In order to obtain the proper bandpass response, the values of the various coils between the cathode of the first color IF and the grid of the second stage, as well as their tuning capacitors must be held very accurately. In fact, C902 is a 10% capacitor, series resonating with L902. Thus a portion of the bandpass response required is obtained by precision, fixed tuned, components while another portion is provided by the high Q, closely coupled, resonant transformer T900.

A rather unique circuit involves the use of a neon lamp E900 in the plate circuit of the second bandpass amplifier stage. This lamp serves as an indicator on the front panel of the color TV receiver to show when a color transmission is received. In the actual circuit of Figure 9.6 it is possible for the technician to misadjust the color killer bias and for the viewer to misadjust the color intensity control in such a way that this neon lamp will not indicate the presence of a color transmission.

Color Demodulators

One of the few circuits which is completely new and unique in color TV receivers and which will not be found in monochrome receivers, FM or broadcast receivers, is the color demodulator. This circuit can be analyzed as a switching circuit, or as a phase comparator circuit, but the basic features can be seen from the functional block diagram of Figure 9.7. The 3.58 mHz color sub-carrier is phase and amplitude modulated. It goes to the demodulator, together with a color synchronizing signal which has a fixed phase and amplitude. In Figure 9.7 we have chosen to show the $-X$ vector and therefore the output of the demodulator will be the $-X$ video signal. The basic color demodulator function can be described by saying that two signals of essentially the same frequency, 3.58 mHz, are fed in and one signal, of a much lower frequency, comes out. The high frequency components are attenuated by means of the LP [Low Pass] filter or, in many circuits, a 3.58 mHz trap is used.

It will be remembered that every color TV set has two identical demodulator circuits whose output differs only according to the reference phase of the color synch signal which is supplied to them. For X and Z demodulators this means that the $-X$ synch signal will be phase shifted 100.9° from the color burst transmitted over the air, and the $-Z$ signal will be phase shifted 62.1° from the $-X$ signal. Where R-Y and B-Y demodulators are used, the color synch signals supplied to them

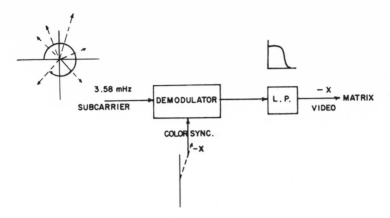

Figure 9.7—COLOR DEMODULATOR PRINCIPLES

will be phase shifted 90° from the reference signal and an additional 90° between each other. It is essential to remember that only these phase differences determine the output of the demodulator and that the same color sub-carrier can be supplied to both the demodulators. We will later consider a special system, used by Motorola and a few other manufacturers, in which the same color synch signal is supplied to both demodulators and the color sub-carrier is phase shifted. The vast majority of receivers, however, use the system shown in the block diagram of Figure 9.1 where the 3.58 mHz color sub-carrier is supplied to both demodulators and different color synch signals are supplied to each demodulator.

One way to analyze the operation of the demodulator is to consider the instantaneous vector relationships between the color sub-carrier and the color synch signal. Figure 9.8 shows the different vectors R which are obtained at various instants as both the burst, or color synch vector, and the sub-carrier vector rotate through a cycle of the 3.58 mHz synch wave.

As the reference vector rotates through one sine wave motion, the sub-carrier vector changes in amplitude and phase angle but is also basically at the same frequency. The sum of the two vectors varies also in amplitude and phase angle; it is this resultant R which constitutes the video X or Z signal. Filtering out the higher frequencies then completes the process and provides the final video signal. In this illustration only a few instantaneous values have been chosen to explain the principle of phase detection. Actually this is a continuous process with rapid variations in phase angle and amplitude.

A different method of explaining the operation of a synchonous demodulator is illustrated by the waveforms of Figure 9.9. The waveforms of the X demodulator and the Z demodulator are shown side by side to illustrate how different outputs are obtained with the same 3.58 mHz

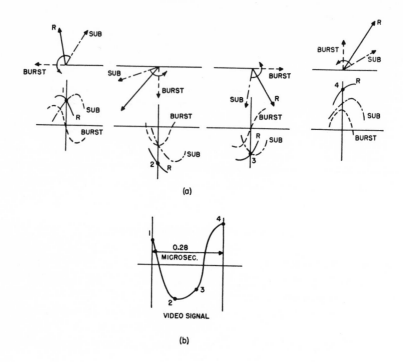

(a)

(b)

Figure 9.8—VECTOR RELATIONS OF A SIGNAL VERSUS REFERENCE

Figure 9.9—DEMODULATOR WAVEFORMS

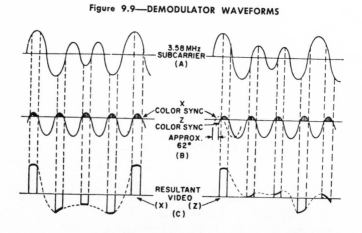

Electronics World

(A) Modulated color subcarrier. (B) 3.58-MHz oscillator
signal. (C) Resultant video for X channel (left) and Z (right).

sub-carrier input. The sub-carrier is allowed to pass through the demodulator tubes only during the brief positive peaks, the shaded areas, of the X color synch signal. As a result, short pulses will appear at the output as illustrated at the bottom line. These short pulses are passed through a low pass filter and then form the video waveform indicated by the dotted line. The Z color synch signal is of the same frequency as the X color synch, but approximately 62° displaced in phase. As a result, the Z demodulator is gated on during different portions of the color sub-carrier and the short pulses allowed to pass through the tube are of different amplitudes and polarities as indicated by the bottom right hand waveform. Again the action of the low pass filter integrates the short pulses into a video signal.

From these explanations and from the block diagram of Figure 9.7 it becomes apparent that the type of circuitry that is required for a demodulator is one that permits a switching or gating action. Figure 9.10 shows a typical pentagrid demodulator circuit. The 6BY6 vacuum tube has the characteristic that its separate suppressor grid has a large effect on the plate current. When the suppressor grid becomes negative, the plate current can be completely cut off. The control grid is biased in such a way as to avoid complete cut-off and, instead, permit amplification of the entire 3.58 mHz color sub-carrier. Assume that at a particular instant the relationship between the sub-carrier and reference voltage is as the example in Figure 9.8.

The total plate current that can flow at that instant will be proportionate to the DC component from the two signals. Similarly at each instant, the sum of the two signals will determine the plate current. Thus the plate current, and subsequently the voltage developed across the plate load, are direct functions of the phase angle and amplitude of the color sub-carrier. It is important that the amplitude of the reference carrier remain constant at all times, since otherwise the incorrect video signal will be produced.

Another way of looking at the operation of the circuit of Figure 9.10 would be to refer to the waveforms of Figure 9.9, and consider the suppressor grid biased near cut-off, so that only the positive portions of the reference signal allow current to flow through the 6BY6 tube. From a practical point of view it is immaterial whether the circuit is analyzed in terms of the switching or gating function or in terms of the addition of vectors since the result will be the same.

In the circuit of Figure 9.10 R1 is the chroma control which corresponds to R902 in Figure 9.6. The cathode resistor R2 is merely there to stabilize the circuit and C1-L1 act as low pass filter to remove all signals above 500 kHz.

As shown in Figure 9.10 the 3.58 mHz reference signal is applied directly to the suppressor grid. In the actual circuit there is a tuned network, resonating at 3.58 mHz and part of the color synchronizing section, which provides a DC return for the suppressor grid. The amplitude

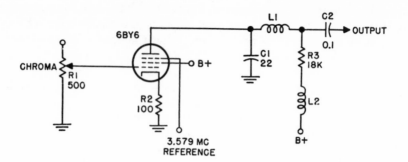

Figure 9.10—PENTAGRID DEMODULATOR CIRCUIT

of the applied reference carrier and the color sub-carrier must be adjusted for the characteristics of the control grid and suppressor grid so that the correct video signal results. In most receivers the reference signal has equal amplitude and a single chroma control adjusts the sub-carrier amplitude for both demodulators. There are a few variations of the circuit of Figure 9.10, mostly in the tube types used and as regards the filter at the output.

A complete color demodulator for both X and Z signals using a special 6GH8A pentode is illustrated in Figure 9.11. In this circuit the color sub-carrier from the bandpass amplifier goes to the color gain control and is then applied to the screen grids of the X and Z demodulator. The resistance values of the voltage divider, consisting of R2, R1 and the color gain potentiometer, are chosen so that the +140 volt B+ is divided down to approximately 1.6 volts. With no signal coming in from the color gain control, this low screen grid voltage will effectively prevent the tubes from passing any current. In this particular circuit the positive portions of the color sub-carrier will therefore control the plate current through the demodulators. The color synch signal itself is applied to the control grids of the two demodulators, but at two different phase angles. Across the secondary of the transformer shown at the lower left, the phase relationship between the reference signal and the Z signal will be approximately 342°. R7 and C6 are used primarily to establish the grid leak bias, and do not substantially change the phase. The X color synch signal must be 62.1° phase shifted with respect to its Z counterpart. This is accomplished by the combination of L2, C3 and R4. Again C2 and R3 merely serve to establish grid leak bias. Both the Z and X demodulator are shown to have the same plate resistance and, since they receive the same sub-carrier signals on the screen grids the average output amplitude at their plates will be the same. It is important to realize that these are averages and not actual, instantaneous signal voltages.

The grid capacitor and L2, respectively L3, serve as the first section of the low pass filter which attenuates the 3.58 mHz component. The

comparison with the waveforms of Figure 9.8 or Figure 9.9 can still be
made in the circuit of Figure 9.11.

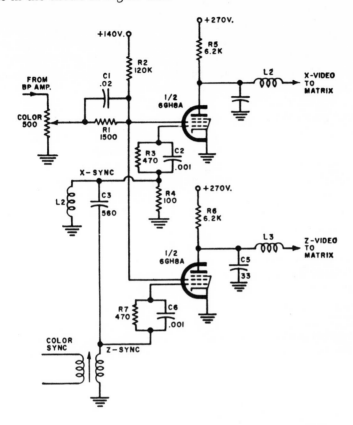

Figure 9.11—SIMPLIFIED DEMODULATORS AS USED IN RCA SETS

We have shown how vacuum tubes can be used, by using the control
grid and the cathode, the control grid and the screen grid or the control
grid and the suppressor grid, to provide the essential gating or phase
comparing action. From these examples it is apparent that a triode or a
transistor circuit can provide the phase comparison function, but it is
not obvious that a two-element device such as a set of diodes can perform the same function. Figure 9.12 shows the diode demodulator circuit used in the General Electric 11-inch portable receivers and in
many other recent TV receiver models. This circuit is called a balanced
demodulator because it depends on waveforms of opposite polarity and
currents of opposite direction. The 3.58 mHz color sub-carrier is applied
to both diodes through capacitor C3 and the color synch signal is
applied across the center-tapped transformer winding at the left. The
color synch at C1 is 180° out of phase with the signal at C2. It is im-

portant to remember this particular feature since it provides the balanced action and without it the circuit would not operate properly.

Neglecting for a moment the demodulator action, let us analyze the effect of diodes D1 and D2 and the three resistors on the sine wave signal applied to C3. D1 and D2 will pass alternate half cycles of the color sub-carrier and a current will flow through the resistors. With R2 adjusted to the exact center of the resistance network the currents flowing through D1 and D2 will cancel each other and no signal will go to the color difference amplifier grid. With the color synch signal applied across the tapped transformer winding, synch voltage waves of opposite polarity will be applied across C1 and C2 and these will be added to the voltages due to the color sub-carrier. If the color synch signal were of the same phase as the sub-carrier at both C1 and C2, the output would still be zero, but because of the center-tapped transformer, this can never be.

One way of illustrating this operation would be to say that the color synch signal, applied in the proper polarity to D1 and D2 respectively, acts to forward bias the diodes in such a way that they can conduct during that portion of the color sub-carrier cycle. In other words, the color synch signal acts as a keying or gating signal which turns diode D1 and D2 on and off. At the right of Figure 9.12 the voltage waveforms are indicated. E_c is the chroma signal applied both across D1 and diode D2 on top of the positive, respectively negative, E_{ref} [The color reference or synch signal]. Note that across the total resistance, R1, R2 and R3, the amplitude due to the current passed by D1 and D2 will be twice the applied signal, the normal effect of a voltage doubler circuit. To turn on D1 a positive color synch cycle is necessary and to turn on D2 the synch must be negative ($+E_{ref}$ and $-E_{ref}$). The output at R2, the center tap, will be approximately half of that and this is based on the vector representation shown at the bottom line.

The diode demodulator shown in Figure 9.12 will produce either the Z or X, or the R-Y or B-Y signals, depending on the phase of the color synch signal that is applied to the center-tapped transformer secondary. In the GE portable color receiver two of these balanced diode demodulators are used to produce the R-Y and the B-Y color difference signal

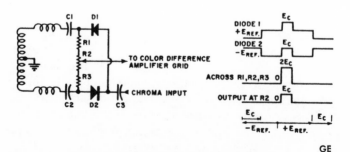

GE

Figure 9.12—BASIC DIODE DEMODULATOR

respectively. A single primary is used to drive two identical center tapped secondaries and the primary determines the actual phase angle of each color synch signal. One of the limitations of the diode demodulators is the fact that, unlike vacuum tubes or transistors, they cannot provide any gain. As a matter of fact, the levels of the sub-carrier and the color synch signal applied to the diode must be of sufficient amplitude to assure operation over the linear portion of the diodes.

A typical circuit using three diode demodulators is shown in Figure 9.13. Although individual components are shown in this circuit, in some receivers a single IC contains all three color demodulators. The 3.58 mHz color reference signal is applied through the phase shifting 220 pf capacitor and the 18 microhenry inductor, both in heavy black lines, to the red and blue demodulators. The green demodulator receives the reference signal without phase shift. All three demodulators are identical and their operation is the same as described for Figure 9.12. In place of

Figure 9.13—TYPICAL DIODE DEMODULATOR CIRCUIT

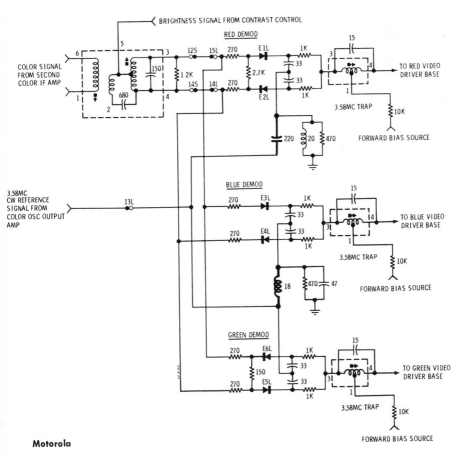

the resistors R1, R2, and R3 in Figure 9.12, the circuit of Figure 9.13 contains two 33 pf capacitors which provide the same balancing effect as the resistors.

Each of the three outputs has a 3.58 mHz trap and is directly coupled to the base of the red, green, and blue driver transistor. One unusual feature of this circuit is that the brightness signal, coming from the contrast control, is added directly, via the center tap of the demodulator transformer. This means that the demodulated video signals will not be the color difference signals but the actual, composite red, green, and blue video signals. The primary of the demodulator transformer and the secondary are both tuned for maximum response at this frequency.

Another type of synchronous demodulator found in many older TV receivers is the so-called gated beam or sheet beam type. The service man will be familiar with the gated beam FM detector circuit using the 6BN6. This circuit is found in many low-cost monochrome TV sets and in some FM broadcast receivers. The principle of the gated beam tube is that, instead of a suppressor grid which controls the plate current somewhat, it contains an electrostatic lens which deflects the stream of electrons going to the plate. This is very similar to the action of the deflection plates in an electrostatic type picture tube. One advantage of this system is the fact that the action of the control grid and of the deflection element are completely independent and that both elements can be designed to have very close control over the plate current, without drawing appreciable power themselves.

Figure 9.14 shows two sheet beam tubes used to provide the three color difference signals directly from the color sub-carrier and the 3.58 mHz color synch signal. The color sub-carrier is applied to the control grid of both tubes, while the color synch signal is applied across two center tapped secondaries, S1 and S2, to the two deflection electrodes D1 and D2, D3 and D4 respectively. Considering first the operation of the B-Y demodulator at the left of Figure 9.14, it is apparent that, depending upon the polarity of the color synch signal, the current from the cathode as modulated by the control grid will go either to plate P1 or to plate P2. The color synch signal applied across S1 must be 180° out of phase with the reference burst in order to produce, when it gates the color sub-carrier, the B-Y signal.

Referring again to the vector diagram of Figure 3.9 we may remember that the B-Y signal is 180° away from the reference signal. The −(B-Y) signal, however, falls exactly into the place of the transmitted reference burst. The circuit of Figure 9.14 can be analyzed when we consider the gating action illustrated in Figure 9.9. The only major difference here is that, because of the center tap of S1 and the action of D1 and D2, one deflection plate will produce a signal corresponding to the positive peaks of the color synch section while the other one will produce a signal corresponding to the negative peaks. Two video signals corresponding

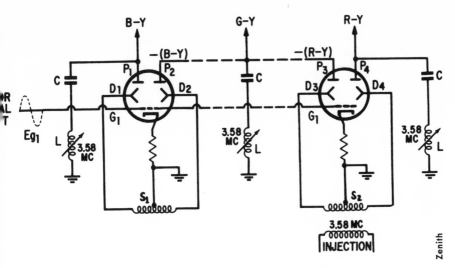

Figure 9.14—BASIC BEAM DEFLECTION DEMODULATOR

to opposite polarities appear at P1 and P2. A series resonant filter consisting of C and L attenuates the 3.58 mHz component and provides the integrating action to make the output of this stage the actual color difference video signal.

The R-Y decoder at the right of Figure 9.14 operates in exactly the same manner except that the color synch signal applied across S2 is 90° out of phase with the color reference signal, corresponding to the R-Y vector. As illustrated in Figure 9.14, the G-Y [Green difference signals] is simply produced by adding the −(B-Y) and the −(R-Y) signals together. Reference to the diagram of Figure 3.9 illustrates that the green difference signal lies in the space between the negative red and negative blue difference signal. The G-Y phase position is such that addition of equal amplitudes of −(R-Y) and −(B-Y) would not result in the correct G-Y. The circuit of Figure 9.14 is greatly simplified and the matrixing resistances have been omitted.

The important thing to understand at this time is the operation, in terms of switching or gating, of the sheet beam color demodulator circuit. In this circuit the color sub-carrier is connected to all control grids in the same manner and separate, center-tapped transformer windings apply the color synch signal to the respective deflection plates. In general this type of circuit is called a high level demodulator because all of the signals are relatively large amplitudes and the output of the respective plates is sufficiently large to drive the grids of the picture tube directly. The sheet beam type of color decoder provides some gain in the tube as well as decoding action.

A widely used Motorola circuit provides the first exception to the rule that the 3.58 mHz color sub-carrier is applied to both demodulators at the same phase and the color synch signal is then supplied to each demodulator at a different phase. As illustrated in Figure 9.15, the color sub-carrier is phase shifted by —45° for the suppressor grid of one pentode and by +45° for the other pentode. At point D resistor R1 and capacitor C1 cause a 45° phase lag. At the suppressor grid E, a 45° lead is obtained by passing the signals through capacitor C2 and then applying it across resistor R2. The net effect of these two phase shifting networks is that the color sub-carrier is applied 90° apart between V1 and V2.

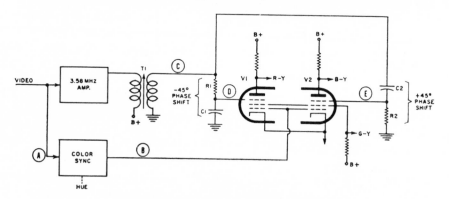

Figure 9.15—SIMPLIFIED MOTOROLA DEMODULATOR

The color synch signal is applied to both tubes at their control grids from point B. Now the circuit of each pentode demodulator can be analyzed in terms of the gating action previously described. As long as the phase relationships of the color synch signal, the sub-carrier and the transmitted color reference burst are correct, the desired R-Y and B-Y signals will be automatically demodulated. To obtain red, the signal at point D in Figure 9.15 must be about 180° out of phase with the color synch signal applied at point B. When this occurs, no current will flow through V1, bringing the plate up to B+. Since the plate is connected to the control grid of the red electron gun of the color picture tube, this will turn the red beam on full. At the same time the sub-carrier at E will be only 90° in phase from the color synch signal at point B, causing V2 to conduct. This lowers the plate voltage of V2 and therefore makes the grid of the blue gun more negative, reducing the blue electron beam current.

The green electron gun operates on the same basis as the red and blue guns which means that, in order to show green, the DC voltage on its control grid must be very close to the B+ value. The G-Y signal in Figure 9.15 is obtained from the combined screen grids of both demodulator pentodes. When green is transmitted, this vector will appear at the suppressor grids of V1 and V2 shifted by + and —45°. This means that both vacuum tubes will draw appreciable amounts of current. With the two plate voltages considerably below the B+ voltage, both the red and blue electron guns will be cut off. Because both plates draw considerable current, very little current is available for the combined screen grids. This means that the potential at the screen grid will be close to the B+ voltage, making the control grid of the green electron gun more positive and resulting in a green screen.

Most of the recent color TV receivers use integrated circuits to perform the color demodulation. In ICs the cost of individual transistors is very low, and, therefore, as shown in Figure 9.16, transistors are used instead of other components wherever possible. In Figure 9.16, the video color difference signals are obtained by demodulating the 3.58 mHz chroma input with the B-Y and R-Y color reference signals.

The basic operation of the IC depends on the use of differential amplifiers. One pair of differential amplifiers, Q15 and Q16, receives the chroma input signal at the two transistor bases. The collectors of these transistors are connected to the two emitters of two pairs of additional, cascaded, differential amplifiers, Q5, Q6, Q7, and Q8, which receive the B-Y reference input at two of the bases. As described previously for vacuum tubes, the demodulating action takes place between the emitter and base of transistors Q5 through 8. These transistors are gated on by the reference signals at their bases. As a result, the output of the collectors of Q5 and Q7 is the B-Y video signal, which is then passed through Q1, providing the emitter-follower output to a driving transistor, not shown in this diagram.

The R-Y demodulator operates in the same manner with the output taken from the collectors of Q9 and Q11. Since the G-Y signal can be

Figure 9.16—DETAILED IC DEMODULATOR CIRCUIT

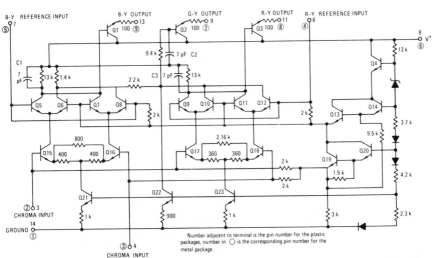

Number adjacent to terminal is the pin number for the plastic packages, number in ◯ is the corresponding pin number for the metal package.

Motorola

made up of the negative values of B-Y and R-Y, the collectors of Q6 and Q8 and Q10 and Q12, respectively, furnish a matrixing signal to the base of Q2 which is the emitter-follower output amplifier for the G-Y signal. At the right of Figure 9.16 a series of transistors, diodes, resistors, and a zener diode are used to assure the correct B+ voltage, and also to act as voltage dividers for the other voltages required on the IC. Three 7 pf capacitors are used to reduce the 3.58 mHz content of the output signals.

A typical application of the integrated circuit shown in detail in Figure 9.16 is illustrated in Figure 9.17. Large capacitors and resonant networks are mounted external to the IC. The three color difference signals go directly to the bases of three drive transistors which are DC coupled to the three picture tube cathodes. The addition of the brightness signal, shown here as luminance signal input, is performed by a single transistor which is effectively in series with the emitters of the three matrixing amplifiers. Individual 250 ohm potentiometers permit adjustment of the color balance.

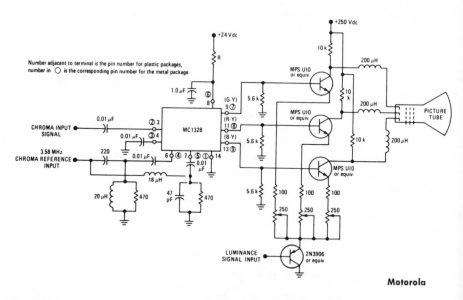

Figure 9.17—IC COLOR DEMODULATOR APPLICATION

The IC-transistor arrangement of Figure 9.17 contains no tuned circuits which might require adjustment and uses less power and space, a great improvement over some of the earlier tube versions of demodulators. Because solid state circuits are used here, the overall reliability of this type of demodulator and matrix amplifier is greatly improved.

Matrix Circuits

From the three equations of Figure 9.2, the red, green, and blue video signals can be made up simply by adding the correct amplitudes and polarities of the Y, X and Z video signals. The matrix section performs this addition by means of simple resistive networks. To understand clearly how these adding circuits work, a simplified network is shown in Figure 9.18. In this network only two voltages are added.

Figure 9.18—SIMPLIFIED MATRIX NETWORK

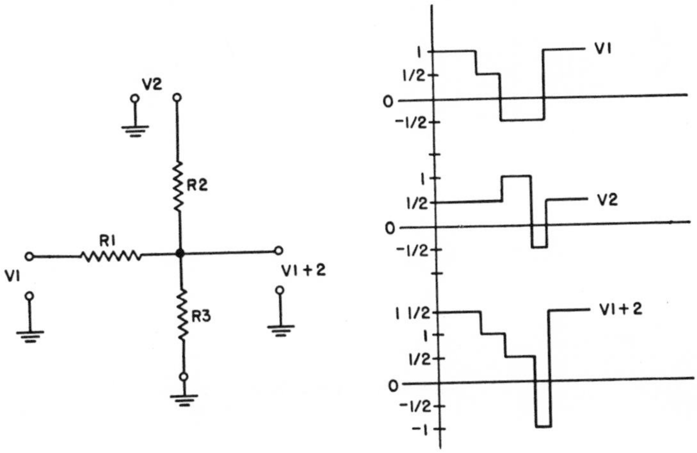

Both voltages appear separately at the two input terminals, but are a single composite signal at the output. The voltages are added instantaneously, positive values adding and negative subtracting. In the event both signals go negative, the result is a negative signal with the two voltages added, just as the arithmetic predicts. The important feature of this circuit is the relative value of R1, R2, and R3. It is necessary that R3 be considerably smaller than both R1 and R2 in order to get good isolation of signals at the inputs. In Figure 9.18 the two series adding resistors R1 and R2 are assumed to be equal, but in the matrixing networks used in color TV receivers the actual resistance values are determined by the matrixing equations shown in Figure 9.2.

Actual matrix circuits in color TV receivers use the same principles but, because of the various vector polarities and amplitudes involved, a typical matrix circuit is somewhat more complex. Figure 9.19 shows, in simplified form, the circuitry necessary to generate the three color difference signals from the X and Z signal. Note that the −X and the −Z are obtained from the demodulator and that each of these signals

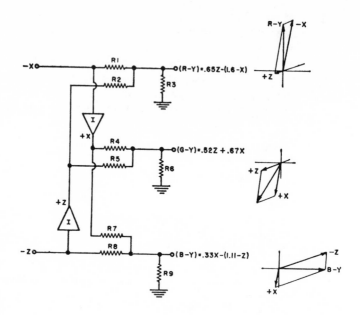

Figure 9.19—PRINCIPLE OF MATRIXING X AND Z

is inverted in polarity by passing it through an inverter stage I. The vector quantities are shown in relative size, together with the numerical relationships which form the R-Y, G-Y and B-Y signals.

From the theoretical circuit of Figure 9.19 we can pass to the actual circuit of Figure 9.20 which shows the matrix and color difference amplifiers used in a very large number of color TV receiver models. One of the outstanding features of this circuit is the common cathode resistor, R15 for all three triodes. Each of the plate load resistors, R3, R9, and R13, are of the same value and the grid resistor arrangement for the R-Y and B-Y amplifiers are also the same. The G-Y signal, as illustrated in Figure 9.19, is composed of positive quantities of the Z and the X signal. Positive values of Z and X signal are obtained at the plate and cathodes of each of the amplifiers. Because the amplitude of +X on the common cathode is less than that required to form the G-Y signal, a small amount of +X signal is fed back from the plate of the R-Y amplifier, via R5, to the control grid of the G-Y amplifier.

A quick comparison between the polarities and the relative values shown in Figure 9.19 and the polarities and relative values of signals

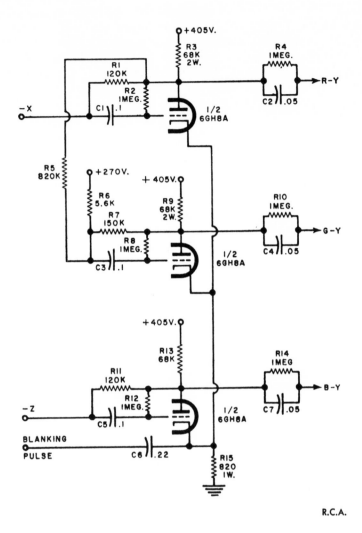

R.C.A.

Figure 9.20—X, Z—COLOR DIFFERENCE MATRIX

which are present at each of the three triodes in Figure 9.20 will show the close correlation between those two matrixing systems. It is essential that the DC components of the R-Y, G-Y and B-Y signal are applied to the control grids of the color picture tube. For this reason each of the leads contains a parallel resistor and capacitor combination from the plate of the color difference amplifier to the control grid.

A complete color demodulator and matrixing circuit as used in a recent Admiral color TV receiver is shown in Figure 9.21. The 6L88 demodulator tube V502 is essentially two pentodes in a single envelope having a common screen grid, control grid and cathode. The color sub-carrier signal is applied to the common control grid and the two color synch signals, in the correct phase arrangement, are applied to the two suppressor grids. The tint control changes the phase of both color synch signals and will therefore affect the hue or tint of all demodulated signals. One plate, pin 6 of V502, goes to plate load resistor R515 and, through the 3.58 mHz trap, via a parallel RC network to the blue electron gun control grid. R523 adjusts a DC level on this control grid and therefore the blue background. The red color difference signal is demodulated from the other plate and again goes through a 3.58 mHz trap and the parallel RC combination to the control grid of the red electron gun. The red background control functions in the same way. For circuit stability a small portion of the red and blue color difference signals are fed back to the control grid by C510 and the voltage divider R520 and R525.

The circuit of Figure 9.21 operates in the same basic fashion as explained for the Motorola system of Figure 9.15, but here the color sub-carrier is applied to the common control grid while the phase shifted synch signals are applied to the two suppressor grids. As in Figure 9.15, the green difference signal is demodulated by the common screen grid. The resistors and capacitors between the two plates and the screen grid and the color picture tube control grids determine the matrixing and the relative DC levels. Although each color difference signal goes through an RC network identical to that described in Figure 9.20, the actual DC level at each electron gun control grid is set by the background control. These three potentiometers determine the background or DC level and also have some effect on the video amplitudes.

In most solid state color TV receivers the matrixing circuit is greatly simplified since the R-Y and B-Y reference signals are used for demodulating. This means that at least these two color difference signals are obtained directly. Matrixing is then only necessary to obtain the G-Y output. The circuit of Figure 9.16 has illustrated how a single IC can provide all three color difference signals. In this circuit only two resistors, a 2.2K and a 9.4K resistor are required to generate the G-Y output from the appropriate polarities of the B-Y and the R-Y signals.

If three separate demodulators are provided, as in Figure 9.13, matrixing is not required at all, but three separate phases of the 3.58 mHz reference signal, must be available. In that particular circuit, even the brightness signal is added directly to all three color difference signals so that the output of the circuit is the actual red, green, and blue video signal which can drive the color picture tube.

Defects in the color demodulator and matrix circuits can usually be recognized by the fact that one or two, or even all three of the colors,

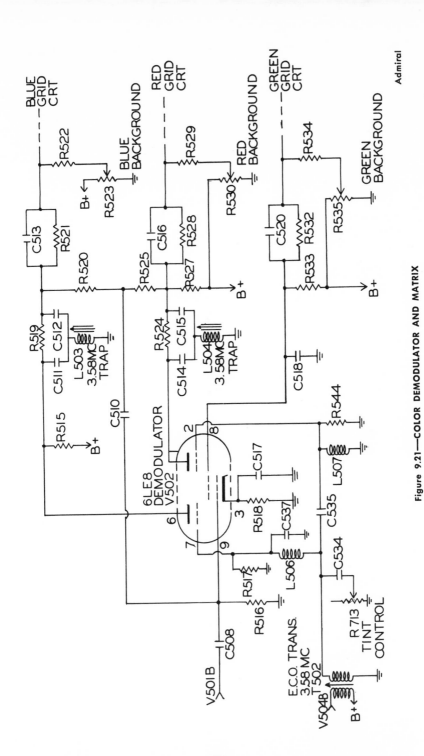

Figure 9.21—COLOR DEMODULATOR AND MATRIX

Admiral

127

will be entirely missing from the screen. If at least one color appears, this means that the demodulators of the other colors may be defective or else that defects may exist in the matrixing or color reference phase shifting circuits. Chapter 19 describes the type of defect that results from malfunctions in the color demodulator and matrixing circuits.

10

COLOR
SYNCHRONIZATION

In the previous chapter the operation of the color decoder section has been discussed; it was pointed out that this section requires a 3.579 mHz reference signal. This reference signal must correspond exactly in phase and frequency to the basic color sub-carrier at the transmitter. For this reason the transmitter sends out short bursts of the 3.579 mHz reference signal. Each of these bursts is timed to coincide with the trailing ledge of the horizontal blanking pedestal. At the receiver this burst must be removed from the blanking ledge and must be expanded into a continuous sine wave signal. There are two basic methods by which a correct 3.579 mHz sine wave signal can be obtained from the color synchronizing burst.

One method involves the use of a 3.579 mHz oscillator which is closely controlled in frequency and phase by an automatic control circuit. With proper adjustment this oscillator then produces a sine wave which is in all respects a duplicate of the transmitter reference signal. The second basic approach to color synchronization makes direct use of the transmitted synchronizing burst. By exciting a resonant circuit with its resonant frequency, continuous oscillations can be produced. Because the Q of any practical resonant circuit is limited, the oscillations so obtained die out after a few cycles. If, however, the Q of the "ringing" circuit is sufficiently high, the oscillations will last until the next color synchronizing burst occurs. This second method of obtaining the color reference signal does away with automatic frequency control systems, but requires amplification and limiting and usually a 3.579 mHz "ringing" crystal.

In some recent TV receivers a combination crystal ringing and oscillator circuit is used which is an oscillator that requires the color synch burst to initiate the oscillation.

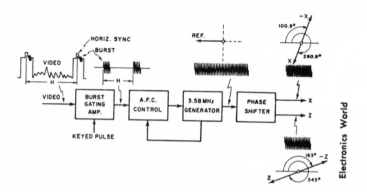

Figure 10.1—COLOR SYNCH FUNCTIONS

In addition to color synch circuits this chapter will deal with the color killer section, because it usually operates in connection with the color synch circuit. Automatic Chroma Control (ACC) is also described here.

The functions of the color synch section can be divided into three separate steps. The first step is to remove the reference burst from its position on the trailing edge of the horizontal blanking pulse. This is generally called color burst gating. In the second step the reference burst is compared with the locally generated signal, except in the crystal ringing circuits where another step of amplification and limiting is necessary. The third step is the generation of the color synch signal itself and the phase shifting to obtain the proper X and Z, or R-Y and B-Y synch signals.

The functions to be performed by the color synch section are illustrated briefly in Figure 10.1. The composite video, including the horizontal pulses and the reference burst, are fed to the burst gating amplifier which is controlled by the horizontal flyback pulse. The output from that stage is the color reference burst only. This burst operates the automatic frequency and phase comparer which, in turn, compares the frequency and phase of the transmitted reference burst with the locally generated signal. At the output of the 3.58 mHz generator a continuous sine wave, corresponding to the original reference signal, is provided. This sine wave is then phase shifted as indicated in Figure 10.1 to provide the X or Z vectors or the R-Y and B-Y synch signals. In the crystal ringing circuits the two boxes labelled "AFC control" and "3.58 mHz generator" will be supplanted by the crystal ringing circuit and its amplifier and limiter stage. The phase shifting is still required.

Color Burst Gating

For either method of color synchronization the color synch burst must be removed from its perch on the horizontal blanking ledge. This is accomplished in the gating section which basically functions just like the gating tube in a keyed AGC circuit. In most color TV receivers the burst gating section consists of an amplifier which is normally cut off and is gated "on" only during the period of the color burst. The gating signal is derived from the horizontal sweep section, usually from a convenient point on the flyback transformer.

A typical burst gating circuit, as used in many transistor type color

Figure 10.2—TYPICAL BURST GATE CIRCUIT

Motorola

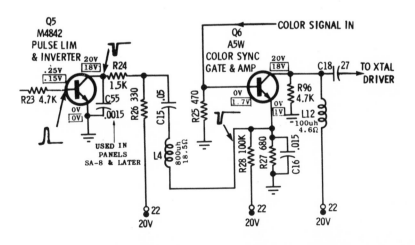

TV receivers, is shown in Figure 10.2. Transistor Q6 is the actual color synch gated amplifier but it obtains its keying pulse signal from Q5, which acts as pulse limiter and inverter. A positive pulse from the horizontal flyback circuit is applied at the base of Q5. This pulse is inverted and limited and appears as negative going pulse at the collector, which goes to a series L-C circuit to shape the pulse for application to the emitter of Q6. The base of Q6 receives the composite color video signal, of which the horizontal synchronizing portion is shown, in a simplified drawing of Figure 10.3A. In order to keep Q6 cut off, the emitter is biased through the voltage divider, R-27 and R-28.

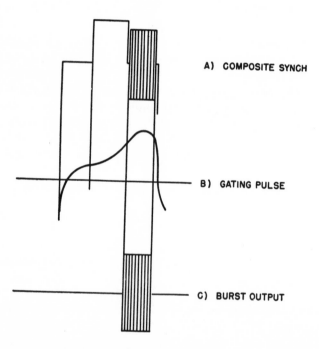

A) COMPOSITE SYNCH

B) GATING PULSE

C) BURST OUTPUT

Figure 10.3—RELATIONS OF SIGNALS AT BASE, EMITTER, AND COLLECTOR

Only when the gating pulse appears at the emitter is this positive bias overcome and Q6 can conduct.

Figure 10.3 shows the relations of the signals at the base, emitter, and collector. To illustrate the gating action, the gating pulse itself is shown as current, rather than voltage, to show that the gating transistor can only conduct during this period. At the collector only the color burst portion is amplified. The important aspect of Figure 10.3 is the wave shape of the keying pulse and its amplitude. Only the tip of the keying pulse is effective so that only the color burst is allowed to pass. If the entire flyback pulse would turn transistor Q6 in Figure 10.2 on, the synchronizing pulse and both slopes of the blanking pedestal would be amplified along. For correct operation both the keying pulse and the video signal itself must be of just the right amplitude and timing. Before attempting to troubleshoot a burst gate amplifier always be sure that the horizontal sweep is in close synchronism. If the picture is unstable horizontally, correct gating action is impossible and as a result the reference frequency supplied to the decoder cannot function properly either. In other words, horizontal synch troubles will invariably cause color trouble as well. Correct the horizontal defects first, then the color portion may be found to be working correctly.

Burst gate amplifiers with essentially the same circuitry as shown in Figure 10.2 are also used when the entire color synchronizing section is contained on a single integrated circuit. In older color TV receivers using vacuum tubes, the burst gating is frequently done in a pentode or similar sharp cut-off tube. The principle is the same as in the transistor circuit, in that the amplifier is biased to cut off, and only when the horizontal flyback pulse appears at the keying element, usually the cathode or the suppressor grid, is the amplifier turned on.

Oscillator and APC Systems

For the serviceman the color synchronizing circuit represents an even greater trouble center than the horizontal system was in early monochrome receivers. The most complicated alignment procedure will be found in receivers using a local 3.579 mHz oscillator and APC system. For this reason a thorough understanding of the circuits and their operation is essential to a successful color service technique.

Figure 10.4 shows a block diagram of a typical phase detector and local oscillator system. The outstanding features of this system are the feedback and the fact that a phase shifting network is employed. Feedback of the locally generated frequency is also required in the conventional horizontal AFC systems. The phase detector, usually a double diode, compares the phase of the burst signal with the locally generated signal. Depending on the difference in phase, an error voltage is generated that is applied to the reactance tube or transistor. This stage is connected in such a way that its reactance appears in parallel with the tank circuit of the local oscillator. The reactance of the reactance stage depends on the bias applied to it and this bias is, in part, the error voltage. Thus the error voltage controls the resonant frequency of the oscillator. When the oscillator is off, error voltage is generated until the correct frequency is reached. The function of the phase shifting network is usually twofold. It supplies the proper phase reference signal to the demodulators and it also determines the phase of the oscillator signal which is fed back to the phase detector.

Figure 10.4—BLOCK DIAGRAM OF COLOR APC

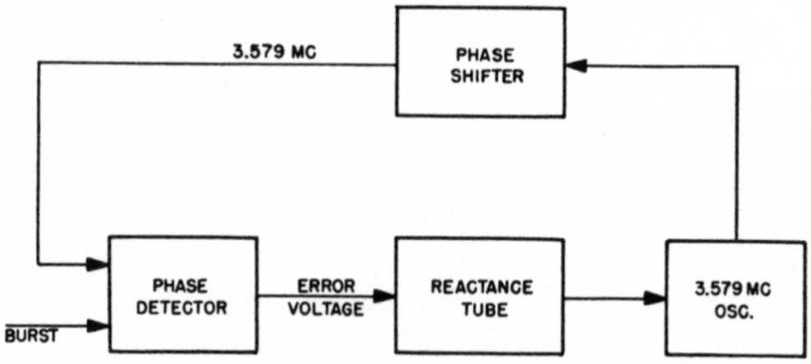

Before going into actual circuitry consider what the phase detector and oscillator must accomplish. They must reproduce a sine wave at 3.579 mHz which is exactly the same in frequency as the 8 cycle burst transmitted at the end of each horizontal line. The phase of the locally generated signal must be within about 10 electrical degrees of the burst phase, the amplitude must remain constant, and no radiation from the oscillator should get into the video amplifier stages. For the phase detector these requirements mean considerable precision, balance, and careful alignment. The oscillator must be tuned almost exactly to the right frequency since the phase detector cannot be expected to "pull in" when the oscillator is off frequency more than a few cycles. With these considerations in mind, the operation of a typical circuit should be studied, noting especially the effects of slight misadjustment of the various controls.

A typical color synch section and its associated color killer circuit is shown in Figure 10.5. Neglecting the killer portion for the moment, the burst gating amplifier V1, at the left, is normally cut off because of the high cathode resistor, 39K. When the horizontal keying pulse makes the control grid positive, the reference burst which occurs at that instant is amplified. T1 is a tuned transformer with its secondary electrically center tapped by the double winding. The primary of T1 is tuned to 3.58 mHz by the plate capacity of V1.

To understand the operation of the phase detector we see that equal amplitudes of the reference burst, but with 180° phase difference, are applied through the two 330 pF capacitors to the plate and cathode of the phase detector respectively. At the same time, the 3.58 mHz oscillator signal is applied to the opposite plate and cathode at one fixed phase. When this signal is exactly in phase with the reference burst, neither of the two diodes can pass current. Depending on the direction of the phase difference between the reference burst and the signal from the oscillator, one or the other of the diode sections will draw current. If the upper diode section draws current, it must flow from the cathode to the plate through R1 and to the control grid of the reactance tube V3. If the phase difference is in the opposite direction, the lower diode will draw current but in the opposite direction, so that the voltage at the control grid of the reactance tube V3 will be of opposite polarity. The voltage on the grid of V3 will therefore be an indication of the phase relationship between the reference burst and the locally generated 3.58 mHz signal. This voltage is usually called the error voltage.

The circuit is similar to the AFC circuit used in some horizontal oscillator sections and resembles the ratio detector used in FM receivers.

The reactance tube V3 is connected across the 3.58 mHz tank circuit L1, C3 so that the capacitance of V3 contributes to the resonant frequency of this circuit. At the grid of V3, C1 and R3-C2 provide a long enough time constant to maintain the error voltage, from one occurrence of the reference burst to the next. This is essential since the error voltage will otherwise fluctuate between reference bursts and this would then vary the reproduced colors at the horizontal sweep frequency.

The major frequency controlling element of the 3.58 mHz oscillator, V4, is the crystal which is cut to resonate at that frequency. Note that the crystal is connected between the control grid and the screen grid of the oscillator tube V4. The oscillating portion of V4 is a triode, consisting of a cathode, the control grid and the screen grid, and isolating the plate circuit as concerns the oscillator frequency. V4 will oscillate at approximately the correct frequency even if V3 and the phase detector are inoperative. To achieve the exact phase of the 3.58 mHz oscillator both the reactance tube and the tank circuit, L1, C3, must be operating correctly. Note that the only adjustment available to the technician is L1, which varies the total resonant frequency of the tank circuit. The phase detector, however, is influenced by two other resonant circuits, T1 and T2. If T1 is misadjusted, the phase of the reference bursts will be incorrect, resulting in an error voltage and in the wrong phase of the oscillator. If T2 is misadjusted, the phase of the signal fed back for comparison will be wrong, resulting in an error voltage and an overall phase shift. Thus, although L1 controls the phase of the oscillator directly, both T1 and T2 must also be adjusted properly to assure that the proper phase of the color reference signals will be generated.

The tint or hue control shifts the phase of the color reference burst slightly as it is applied to the phase detector. C3, a 120 pF capacitor, is connected from one side of the T1 secondary to the tint potentiometer. This RC circuit has the effect of unbalancing the secondary of T1 and thereby shifting the phase of the color reference burst as it arrives at the detector. In most receivers this control is mounted on the front panel and has a range of approximately ± 60 electrical degrees of phase shift. This is usually sufficient to allow the pink portions of the picture to turn green and vice versa.

The circuits associated with the color killer bias generation are also shown in Figure 10.5 because they obtain their signals from the color synch section but their detailed operation will be discussed at the end of this chapter.

A transistor version of the circuit described in detail above is shown in Figure 10.6. Beginning at the left, two solid-state diodes each are used as phase detector and color killer detector. The output of the color killer detector goes to the color killer amplifier stage, not shown here, through capacitor C680. The error signal output of the color phase detector is obtained at the center of the two resistors, R-768 and R-769. A DC amplifier, consisting of two transistor stages contained in a single case amplifies the error signal.

The color synch burst is applied through the center tapped transformer and the other input is feed-back from the 3.58 mHz output transformer to the center of the two diodes. This point is also DC connected to the collector of the phase control amplifier, which controls, by adjustment of the R776, the hue or tint of the ultimate picture.

As in the tube version of Figure 10.5, the DC error voltage controls a variable reactance. In Figure 10.6, the DC voltage amplified by Q640

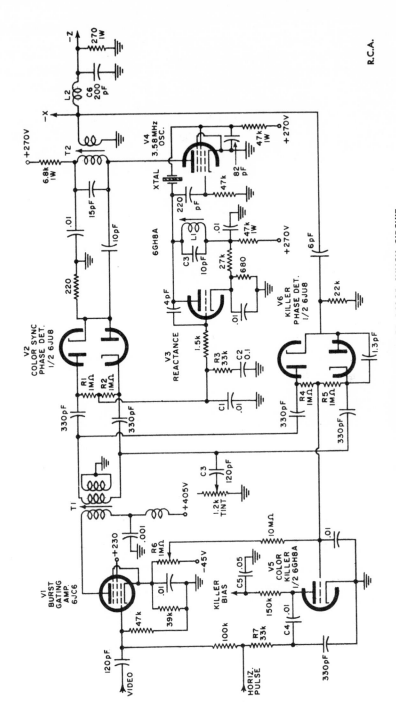

Figure 10.5—DETAILED OSCILLATOR TYPE COLOR SYNCH CIRCUIT

137

controls the reactance diode SC 618 and therefore the frequency of the 3.58 mHz oscillator, much in the same manner as reactance tube V3 in Figure 10.5. Again a 3.5 mHz crystal is used for stability. An extra stage of output amplification is used in Figure 10.6 and the final color reference signal is then obtained from the T606 transformer secondary.

Although the number of tubes in Figure 10.5 is less than the number of solid-state components in Figure 10.6, the transistor color APC circuit requires much less power and is much more stable than its vacuum tube cousin.

The alignment and troubleshooting of color synchronizing circuits is treated in Chapter 15.

Crystal Ringing Systems

Completely different in operation from the oscillator APC system is the crystal ringing circuit used in a few early color TV receivers. As mentioned before, in this system the original continuous sine wave reference signal is restored, and, if this is accomplished, no color synchronizing error can occur. The reference signal at the transmitter is a continuous sine wave of constant amplitude and the bursts of syncronizing information sent out are merely sections from this continuous signal. Furthermore, each burst starts and stops at the same point on the sine wave. This feature is important because in the crystal ringing system the start of the next burst must coincide perfectly with the damped oscillation due to the crystal.

Figure 10.7 shows in block diagram form how a crystal ringing circuit operates. The burst gate removes everything except the color synchronizing burst. This burst is applied to a very high Q resonant circuit, a quartz crystal, tuned to the burst frequency of 3.579 mHz. Excited by the burst, the crystal generates damped waves at its natural frequency. These damped waves are amplified and clipped to produce a constant amplitude, continuous sine wave signal. This signal is phase shifted to form the synch signals. It is evident from these principles that the crystal ringing system of color synchronization is considerably simpler to service and align than the other system.

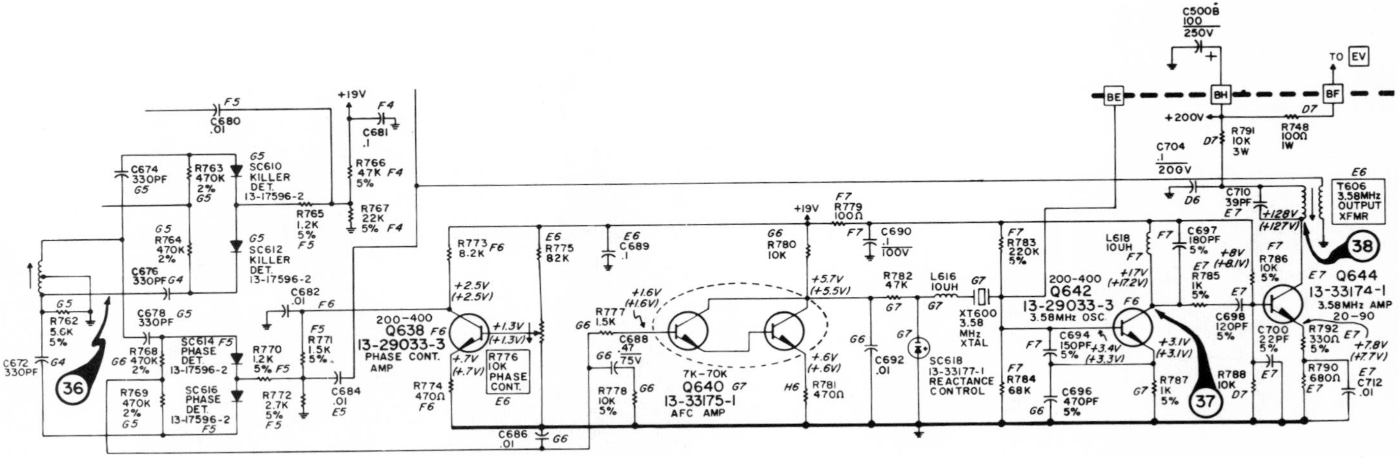

Figure 10.6—TRANSISTOR COLOR APC CIRCUIT

Sylvania

139

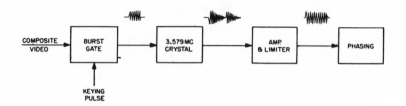

Figure 10.7—BLOCK DIAGRAM OF CRYSTAL RINGING CIRCUIT

Some indication of the operation of the crystal ringing circuit can be obtained from the oscilloscope photographs of Figures 10.8 and 10.9. In the first photograph the large starting amplitude and the exponential decrease of the ringing circuit are clearly shown. In Figure 10.9 relatively constant amplitude exists except for the short increase during the actual reference burst.

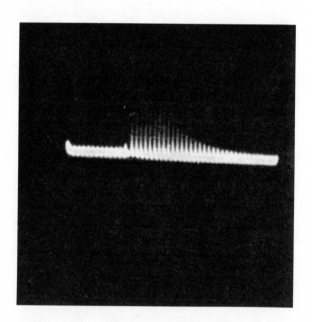

Figure 10.8—DAMPED WAVES AT CRYSTAL OUTPUT

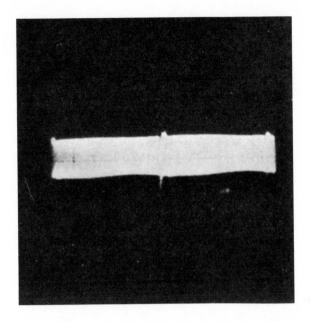

Figure 10.9—COLOR SYNCH AFTER LIMITING

In most color TV receivers designed after 1968, a combination crystal ringing and free-running oscillator system is used. This approach combines the advantage of the crystal ringing method with the stability of a continuous oscillator. Transistor Q7 in Figure 10.10, receives the separated color synch burst and amplifies this to excite the 3.58 mHz crystal. When the resulting output is amplified by Q8, the relatively continuous signal is used as evidence that a color transmission takes place and is coupled through C12 to the ACC and color killer circuits. A modified Colpitts oscillator, Q10, operates at approximately 3.58 mHz. When a color signal from the ringing crystal is received, the otherwise free-running oscillator is locked-in to the correct reference frequency and phase. From the emitter circuit of Q10, the color reference signal then goes to another stage where the correct reference phases to the color demodulator are generated.

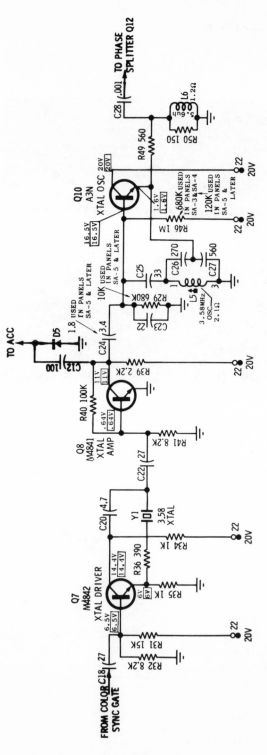

Figure 10.10—XTAL RINGING TYPE COLOR SYNCH CIRCUIT

Motorola

142

One of the advantages of the combination crystal ringing and free-running oscillator circuit of Figure 10.10, is the fact that only one resonant circuit, L5, requires adjustment. Another advantage is the fact that, unlike the phase detector type circuits, there is no time delay or lag until synchronism is achieved. As long as the crystal ringing period lasts, the oscillator will be synchronized. It is important to remember, when aligning this type of circuit, that the oscillator coil can only be adjusted correctly in the presence of a color burst signal.

IC Color Synchronizing Circuits

In many late model color TV receivers, the color synch section as well as the automatic color control section, and in some instances even the color killer section, are contained on a single IC. The circuit illustrated in Figure 10.11 is typical of this arrangement. As in the vacuum tube version of Figure 10.5 and the transistor version of Figure 10.6, the chroma input signal goes to the automatic phase detector (APC) and, separately, to the automatic chroma control (ACC) detector. Each detector also receives a feed-back signal from the oscillator and is keyed by the horizontal keying pulse amplifier. The output of the APC detector controls the crystal controlled circuit. The balanced output of the ACC detector goes to a separate ACC circuit which is part of the color bandpass amplifier section located on another IC. Separate 20K potentiometers are provided for balancing the input to the APC and the ACC detector. The hue control acts on the oscillator output amplifier which has two separate, opposite phase, output terminals. The IC also contains a shunt regulator and bias circuit to assure that the correct B+ voltages are applied within the IC.

In addition to the advantages inherent in using a single IC to perform all of these functions, it will be noted that the only resonant or tuned network is the crystal, which requires no adjustment. This makes the alignment of the color synchronizing shown in Figure 10.11 much simpler than some of the earlier vacuum tube versions.

Interference

One of the problems encountered in the color synchronizing section is the possibility that the locally generated 3.579 mHz signal may interfere with some monochrome receiver or even beat with the video signal or monochrome reception. When the receiver is tuned to a color telecast and the color synch section is properly locked in, there can

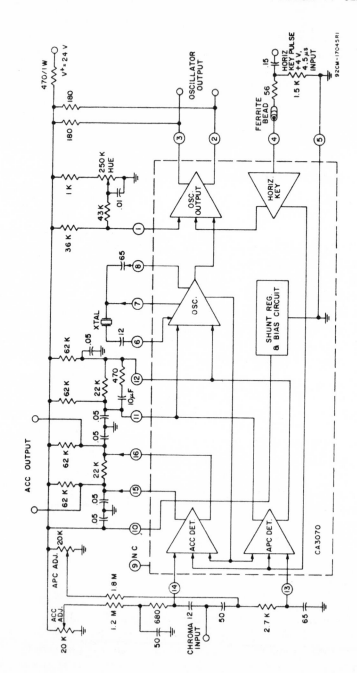

Figure 10.11—TYPICAL IC COLOR SYNCH CIRCUIT

R.C.A.

be no interference within the receiver. It is, however, possible that the set might interfere with another set which is tuned to a different station and receiving monochrome. Most color receivers have sufficient shielding in the color synchronizing section to avoid this radiation, but occasionally freak conditions exist where such interference is observed. The remedy lies in using tube shields and a chassis bottom plate on the color set.

Some early color receivers had a color-monochrome switch to disable the color synch section during monochrome reception. The most frequent source of this type of interference is in receivers using a color killer circuit and a locally generated reference signal. The remedy is to check the color killer operation since that circuit is supposed to keep the 3.579 mHz signal out of the screen. Checking for poor grounds, floating shields, and the like might also show the source of this interference.

Color Killer and ACC Circuits

The function of the color killer circuit is to make sure that spurious signals, noise impulses, and occasional interference cannot be interpreted as color signals during monochrome transmission. If the color killer circuit is inoperative or misadjusted, colored flashes can appear in the monochrome picture. One method of judging the performance of the color killer circuit is to watch a slightly noisy monochrome picture, make sure that no color flashes occur, and then tune to a color transmission and make sure that full color is available.

Referring back to Figure 10.5 we note that the color killer phase detector is connected across the secondary of T1 by means of two capacitors. This phase detector operates in the same basic manner as the color synch phase detector described earlier in this chapter. When the reference signal is absent, during monochrome transmission, the bias on the color killer stage will be less negative and it will conduct during the horizontal pulse. This bias is then used to cut off the chroma amplifier.

A combination color killer and automatic chroma control (ACC) is shown in Figure 10.12. The 3.58 mHz signal from the color reference circuit, such as the crystal ringing signal from the circuit of Figure 10.10, is applied to the base of Q707, the killer detector transistor. The

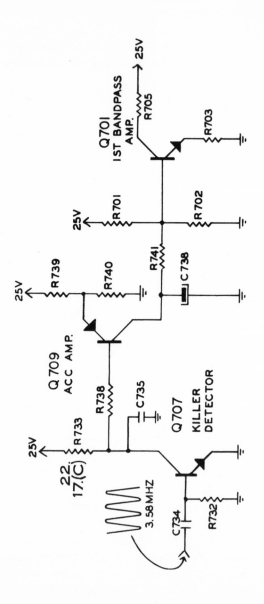

Figure 10.12—COLOR KILLER AND ACC CIRCUIT

Admiral

collector of this stage is direct-coupled to the ACC amplifier, which, in turn, has its collector connected to the base of the first bandpass amplifier, Q701. The AC chroma input to Q701 is not shown in Figure 10.12 but consists of a capacitor going to its base. When no synchronizing bursts are received, during monochrome transmission, this will cause Q701 to be biased beyond cut-off. When color synch bursts are received, beyond a minimum amplitude, the filtering action of C735 and C738 generate an amplified DC voltage which controls the gain of Q701. Q709 effectively acts as DC inverting amplifier. As the amplitude of the chroma signal varies, the gain is held constant by this action.

Among the various defects which can cause defective color killer operation the most likely is, of course, failure of the tube or transistor. If the tube is good, the killer bias itself should be measured with various settings of the killer bias control. It is important that this measurement be made on a monochrome transmission and then repeated on a color transmission. If no color killer bias variation is obtained, the most likely defect is absence of the horizontal flyback pulse and that can be checked by connecting the oscilloscope to the color killer stage. If the horizontal flyback pulse of the proper amplitude appears, DC measurements should reveal any defective component in that circuit. Finally the operation of the phase detector can be checked by tracing the 3.58 mHz signals with the oscilloscope.

The color killer bias is always controlled by a potentiometer, located with the other service adjustments and intended for use by the technician only. During monochrome transmission the potentiometer is first set to the extreme setting which will allow color flashes to appear. It is then turned until a clear monochrome picture is obtained. Next, a color broadcast, or a test signal from a color bar generator, is tuned in and this may then require readjustment of the color killer control to assure proper color reproduction. A recheck on monochrome transmission is advisable.

To sum up the operation of the color killer circuit, we know that it needs the presence of a horizontal keying pulse, the color synch burst and the locally generated oscillator signal or else a bias from the 3.58 mHz oscillator. If no color synch burst appears, the color killer will generate a large negative bias which is usually applied to the sub-carrier amplifier and which will cut off that stage. Defects in the color killer circuit result either in loss of color during color transmission or in the appearance of color flashes during black and white transmission. In either event, the defect can be isolated rather simply by the troubleshooting procedure described above.

11

DEFLECTION
AND SPECIAL
COLOR CIRCUITS

This chapter deals with special circuits which are not found in monochrome receivers and which are unique to color TV set operation. The horizontal deflection system, including the high voltage flyback portion, is basically the same as in monochrome receivers but some additional features are required for color operation. This chapter includes the centering and the high voltage regulation circuits as well as the circuits which are used to correct for the pin cushioning effect, a feature which requires both vertical and horizontal sweep signals. The degaussing systems for the color picture tube and the delay lines used in color TV sets are also described in this chapter.

Centering

In most monochrome TV receivers centering of the picture was accomplished either by a centering magnet on the neck of the picture tube or by varying the tilt of the magnetic focus assembly. In color TV receivers it is much more important to center the picture correctly and accurately because the purity and convergence adjustments described in Chapters 5 and 6 will be greatly affected by the centering. As we have seen from Chapter 5, there is very little room on the neck of the color picture tube, and the addition of a magnetic centering device could interfere with the magnetic purity device and with the operation of the convergence assembly as well. For this reason, practically all earlier color TV receivers use DC centering. By this we mean that a small DC is superimposed on the sawtooth deflection signal, through the horizontal and vertical deflection coils.

Figure 11.1 shows the typical vertical centering circuit in which the amount of DC through the deflection yoke is controlled by a potentiometer across a portion of the vertical output transformer. The exact arrangement of the transformer windings vary between different manufacturers and are similar in the horizontal deflection circuit. The principles of the circuit, however, are the same wherever DC centering is used. The DC going to the vertical or horizontal output transistor or tube is used, in part, to determine the overall vertical or horizontal centering of the picture on the screen. If an open circuit occurs in the potentiometer, there will be no DC flowing through the deflection yoke and the picture will again be at either side or up or down. An open circuit in the centering control usually also reduces the sawtooth

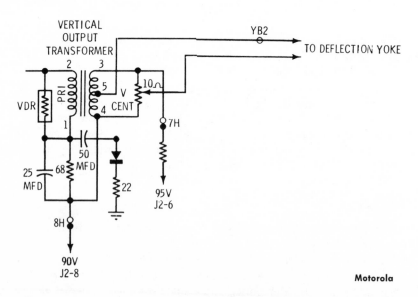

Motorola

Figure 11.1—VERTICAL DC CENTERING

current in the deflection coils, and this, in turn, produces a picture that is either too narrow or too short in height. An open capacitor or open rectifiers would also reduce picture size and would make the potentiometer act as a size control rather than a centering control. In either of these instances simple ohmmeter tests, with the power turned off, will quickly reveal which component has become defective.

High Voltage Regulation

In Chapter 6 the high voltage power supplies of color TV receivers have been discussed to some extent and the need for regulation of the high voltage has been demonstrated. Regulation for the high voltage was provided in earlier models by a special high voltage regulator tube. Some of the more recent color TV receivers provide the high voltage regulation by means of a special circuit in the horizontal flyback section which, however, does not work on the high voltage directly as described in connection with Figure 6.2.

The circuit of Figure 11.2 does not use a special high voltage regulator tube, but high voltage regulation is provided in the output amplifier stage. While this circuit really serves two purposes, we shall ignore the presence of the vertical output signal appearing at the left and concentrate on the operation of the high voltage regulator. A separate winding on the horizontal flyback transformer provides a horizontal pulse which is rectified by diode D1. The rectified horizontal pulses are filtered by the various resistors and capacitors and result in a negative bias applied to the control grid of the horizontal output amplifier. The actual value of this negative bias is determined by the setting of the 50 K bias control potentiometer which provides the DC return path of the diode, D1.

If excessive current is drawn through the high voltage rectifier, not shown in Figure 11.2, this loads down the flyback transformer and reduces the amplitude of the pulses applied to diode D1. Reduced pulse amplitude means a reduced negative bias on the control grid of the horizontal output amplifier. This increases the gain of that stage and thereby compensates for the extra load put on the circuit by the increased current drain in the high voltage section.

In typical operation the bias control is set for an average brightness, monochrome picture, because this will require more ultor current than the average color picture. To make sure that the high voltage is correct, it is advisable to measure its value on a good black and white

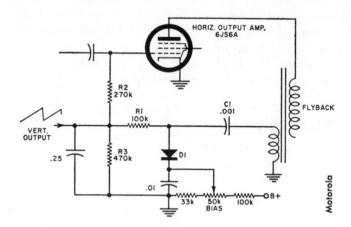

Figure 11.2—H.V. REGULATION AND PIN CUSHION CORRECTION

picture or else the appearance of the picture itself can often be used to judge correct setting of the control. Once the bias control is set, it ordinarily does not require readjustment, unless the horizontal output amplifier or one of the other tubes in the horizontal flyback section has been changed or a component has been replaced. Most of the defects in the circuit result in a complete or partial loss of horizontal deflection and high voltage. To determine which component is defective it is usually necessary to measure the high voltage itself and then measure the grid bias and the voltages on the bias control potentiometer in the regulator circuit shown in Figure 11.2. The use of the vertical sawtooth voltage in connection with this particular circuit will be taken up later in this chapter under the heading of "Pin Cushion."

A high voltage regulating circuit which uses a separate triode as a regulator tube is shown in Figure 11.3. The operation of this circuit again depends on a special winding on the flyback transformer which provides horizontal flyback pulses, the amplitude of which depends on the ultor current. Again the grid bias of the horizontal output stage is varied to control the gain of that tube. This is accomplished by a special circuit using a triode section as pulsed DC control amplifier. The plate of this triode receives the positive horizontal flyback pulses so that it conducts during the flyback pulse period. A small portion of this positive flyback pulse is applied to the control grid through the high voltage regulating potentiometer R6 and a resistance divider network. The

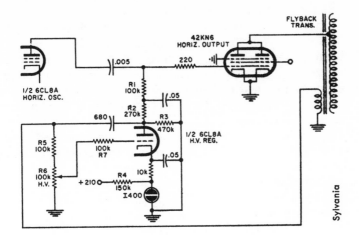

Figure 11.3—H.V. REGULATION TRIODE CIRCUIT

cathode of the regulator is kept at a fixed potential through the voltage divider and the neon indicator tube I400. The cathode voltage will be approximately 52 volts, the normal neon breakdown potential.

The positive flyback pulses going to the plate of the regulator tube cause that tube to conduct and pass current through R3 which results in a negative voltage set up across R3. When the load on the flyback transformer increases due to excess current drawn by the high voltage supply, the amplitude of a positive pulse applied to the grid and plate of the high voltage regulator is reduced and this, in turn, makes the bias on the horizontal output tube less negative, allowing that tube to provide more amplification. Trouble-shooting of this circuit is essentially the same as described above for Figure 11.2. Manufacturer's data for correct voltages should be compared to the measured values.

Pin Cushion Effects

In early monochrome TV receivers, particularly those using optical projection methods, pin cushioning was a frequent trouble. The pin cushion appearance of the raster was usually due to the deflection yoke but, where wide deflection angles and relatively short deflection yokes had to be used, pin cushioning was an inherent phenomenon which could only be eliminated by changing the geometry of the deflection yoke and the tube. In modern color TV receivers the short picture tubes with short deflection yokes are important to keep cabinet size down and pin cushioning generally appears.

Fortunately, pin cushioning can be eliminated by electronic means. Figure 11.4 shows, at the left, the appearance of typical vertical pin cushioning. At the right of Figure 11.4 the current wave form for one vertical scan is shown with the correction currents to eliminate pin cushioning. We see that a different amount of correction current is needed for each of the horizontal lines, with no correction needed at the center of the picture and the maximum correction required at the top and bottom. This current wave form immediately suggests a relatively simple approach to pin cushion correction. We need only apply portions of the horizontal deflection signal to the vertical scan to take care of vertical pin cushioning and a portion of the vertical deflection signal to the horizontal scan to correct horizontal pin cushioning. This is exactly the technique used in modern color TV receivers to eliminate the pin cushion effect.

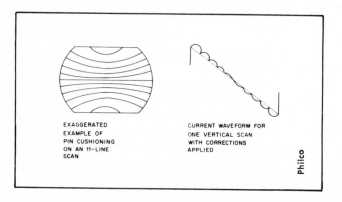

EXAGGERATED EXAMPLE OF PIN CUSHIONING ON AN 11-LINE SCAN

CURRENT WAVEFORM FOR ONE VERTICAL SCAN WITH CORRECTIONS APPLIED

Philco

Figure 11.4—PRINCIPLE OF PIN CUSHION CORRECTION

Figure 11.5 illustrates a transistor circuit permitting separate pin cushion control at the bottom and the top of the raster. Transformer T1 is in series with the vertical deflection coils and couples the correction signal directly to them. This correction signal is obtained by connecting the horizontal flyback pulses to the primary of T1 as well as

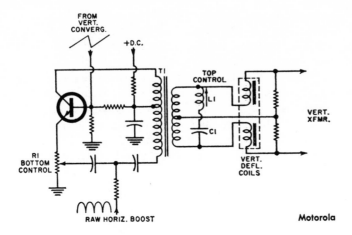

Figure 11.5—VERTICAL PIN CUSHION CIRCUIT

to R1, the emitter resistor of the transistor amplifier. A small vertical sawtooth voltage is applied at the base of this transistor to provide the center reference voltage for the horizontal correction signal. Adjustment of R1 determines the amount of horizontal signal amplified by the transistor. Adjustment of L1 determines the tuning of T1 secondary and thereby the relative amplitudes of the horizontal pulses superimposed on the vertical sweep signal. In actual practice R1 is adjusted to minimize pin cushioning at the bottom of the picture and L1 is tuned to minimize pin cushioning at the top.

Figure 11.6 shows the dynamic pin cushioning correction circuit for an earlier Philco receiver. In this circuit a portion of the horizontal deflection yoke is connected in series with a special transformer, consisting of three separate windings. A voltage dependent resistor, VDR, is connected across the winding to stabilize the total horizontal amplitude. The vertical deflection yoke is also in series with this transformer, but in such a way, that a portion of the horizontal signal is coupled into the vertical yoke and a portion of the vertical signal is coupled into the winding in series with the horizontal deflection yoke. A potentiometer "Pin Cushion Amplitude," is connected between the two vertical coils, and a tuned transformer, labeled "Pin Cushion Phase," is connected between two of the windings of the special transformer and the B+.

A detailed analysis of this circuit would be quite complex since it depends on the magnetic characteristics of the transformer. In effect, however, the "Pin Cushion Phase" adjustment determines the position of the horizontal pulses on the vertical sawtooth, while the "Pin Cushion Amplitude" adjustment controls the amount of horizontal signal which is superimposed on the vertical deflection yoke. The amount of vertical signal coupled into the horizontal deflection yoke is limited by the voltage dependent resistor, VDR, and is relatively stable. The circuit of Figure 11.6 has the advantage of not requiring transistors or vacuum tubes but has the limitation that any defect in the special three section transformer will require an exact and relatively costly replacement part from the manufacturer.

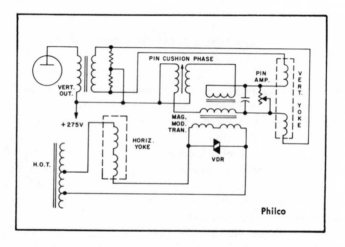

Figure 11.6—COMPLETE DYNAMIC PIN CUSHION CORRECTION

Degaussing Systems

In Chapter 5 we mentioned that early color TV receivers required a field neutralizing magnet ring to eliminate the effects of stray magnetic fields. In those receivers, one of the frequent tasks of the service man was to degauss (demagnetize) the screen area of the color picture tube. Most of the color TV receivers marketed since 1964 are equipped with built-in degaussing coils which perform this operation either automatically or when a switch is actuated.

A fully automatic degaussing circuit is shown in Figure 11.7 and this is representative of most color TV models. Here the degaussing action is provided by the 60 Hz AC line voltage. When the color TV set is first turned on, the thermistor, T, presents a relatively high resistance, causing most of the AC current to flow through resistor R and through the degaussing coil. As the set warms up, thermistor T is reduced in its resistance while the variable resistor, R, increases. This causes the current in the degaussing coil to become smaller and smaller until, when the set is properly warmed up, practically all of the current passes through the thermistor, T, and none of it through the degaussing coil. The characteristics of R and T in Figure 11.7 determine the timing of the automatic degaussing cycle, which occurs whenever the TV set is turned on from a cold start.

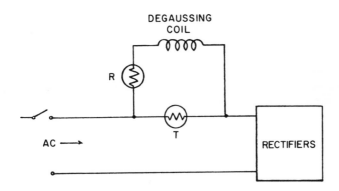

Figure 11.7—AUTOMATIC DEGAUSSING PRINCIPLES

Some color TV receivers have somewhat different degaussing circuits but the principle is the same in all of them. In each case a short burst of AC is applied to the degaussing coil and this then decays rapidly to a steady state. Defects in the degaussing circuit are quite rare. If it seems very difficult or impossible to obtain good color purity, it may be worth-while to check the operation of the degaussing circuit. Simple ohmmeter checks for continuity of the various resistors and the coil itself will be sufficent to locate the defect.

Delay Lines

As mentioned in Chapters 7 and 9, a new component is used in color TV receivers which the serviceman has not encountered in monochrome service work. This item is called a delay line or delay network and its purpose is to delay some of the video signals for a fixed time. While the chrominance signals, X and Z or R-Y and B-Y, are being demodulated, amplified, inverted, and finally appear at the matrix, the brightness signal receives only one or two stages of amplification and then is ready for the matrixing operation. It is apparent that the time required for the chrominance signals to travel from the first video amplifier to the matrix is slightly longer than the brightness signal needs to meet the chrominance portion at the matrix. In order to make sure that the two different signals coincide exactly, a certain fixed amount of time delay is introduced in the brightness channel. In other words, the Y signal is delayed by the same amount of time that the chrominance signal needs to reach the matrix. This delay time is 0.9 microseconds in most systems.

Before analyzing the operation of a delay line, consider the effect of a video signal traveling over a long piece of coaxial cable. Assuming a perfect cable with no losses, the signal will still need some time to travel the length of the cable. The actual travel time for the signal depends on the propagation characteristic of the insulator used and is always less than the speed of light which a signal would have in

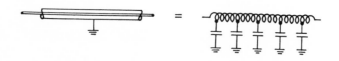

Figure 11.8—EQUIVALENT CIRCUIT OF CABLE

free space. Now consider the electrical equivalent of a coaxial cable as shown in Figure 11.8. The inner conductor has inductance which is in series with the path of the signal. A distributed capacity exists between the inner and outer conductor which is in shunt with the signal. In other words, the cable appears as a distributed series inductance and distributed shunt capacity as shown in Figure 11.8. The time delay due to this cable can be varied simply by cutting off or adding pieces.

In some early color TV sets the delay lines actually consisted of lengths of special cable and in many pieces of studio equipment the cable type delay line is utilized. Such a cable usually consists of a ferrite core with the center conductor coiled around it and a layer of insulation covered by the outer shield. The object is to get the maximum inductance and capacity into a relatively short stretch of cable.

In all modern color receivers the distributed inductance and capacity of the cable are replaced by lumped inductance and capacity, arranged into a fixed delay line. The design of this line is such that, by using several coils and capacitors a constant impedance level is maintained with a frequency response which is flat to at least 4 mHz. Since the service technician need not be familiar with the design criteria of delay lines, only the troubleshooting aspects of these devices will be discussed here.

There will never be any need for adjusting delay lines, but it is possible that repairs or replacements will be needed. If one of the coils opens, no signal will reach the brightness amplifier and completely wrong and blurred colors will result, while no picture at all will appear on monochrome reception. A simple resistance check will determine this kind of defect, but usually such a break occurs inside a coil, permitting no repair but requiring complete replacement. Another type of defect is the shorting of one of the capacitors or grounding of the center wire. This again, will result in loss of brightness signal with the same symptoms as outlined above. Again a simple resistance check will show this type of defect and locate the guilty component.

When one of the coils is shorted or one of the capacitors opens, the result will be only a change in the time delay, and a change in the coincidence of brightness and color. This type of defect is very difficult to diagnose since a number of other causes can produce the same symptoms. The appearance of such a defect may be a very slight misregistration of the three colors at the screen. At the misregistration points the colors will be radically different from the correct ones, but these "wrong" areas may be very small. Poor DC convergence sometimes gives the same effect.

As a rule, shorted coils or open capacitors in the delay line are quite rare and should only be considered as possible trouble spots if none of the other sections yield any defects or misadjustments. The simplest method of determining if a delay line is defective in the manner just mentioned is to substitute a known good one temporarily and compare the picture quality.

12

INSTALLATION

The first eleven chapters of this book cover the theory and operation of a color TV receiver. This basic knowledge is necessary if the servicing of color receivers is to be done in a professional manner. The men who repaired home radios years ago often knew very little about the electronic principles of the equipment they serviced. Testing tubes, replacing suspected components, and taking care of obvious mechanical defects was usually sufficient to keep a man in business. The advent of television quickly showed that the screwdriver mechanic could not operate as efficiently as his more skilled competitor. Color television work requires even more knowledge and technical understanding and leaves less room for semi-skilled operators than monochrome servicing.

The first contact the customer has with the serviceman occurs usually at the time of installation. First impressions being as important as the proverbs say, the serviceman should be sure to appear at his best. Passing over the obvious matter of personal appearance, the next paragraphs show that this installation is not a haphazard affair, but a well-planned, skilled routine. All needed tools should be brought along as well as any required hardware and other equipment. No customer complains about an overequipped installation team or truck, but underequipped TV installers are as familiar as the legendary plumber who always leaves the right size wrench at home. To help

the reader plan an installation, here is a table of tools usually needed
for this job. It would be a good idea to check the truck or car each
morning for such a complement.

TOOLS REQUIRED FOR TV INSTALLATION

Aligning tool (RF & IF coil)
Assorted rawl plugs or anchor bolts
Assorted nuts, bolts, woodscrews, and nails
Assorted sandpaper
Brace and bit (up to ½ inch)
Carpenters hammer
Cleaning fluid
Cold chisel (½ inch)
Concrete drill (¼ to ⅜ inch)
Diagonal cutting pliers
Drop cloth
Electric drill and set of drills
Files, rough and fine
Friction tape
Furniture polish
Gas pliers
Glyptol or service Cement
Guy wire
Hacksaw
Keyhole saw
Long-nose pliers
Lubricant—oil or graphite
Pocket knife
Polyethylene tape
Rubber tape
Screwdrivers, assorted
Soldering iron and solder
Steel nails
Spatula
Steel strap punched
Transmission line
Transmission line standoffs
Turnbuckles, hooks and eyes
Weatherproofing compound
Wire stripper and assorted wire

In addition to the mechanical tools listed above, experienced servicemen also like to take along a portable monochrome TV set, a multimeter and, if possible, a color bar generator. The portable TV set
comes in very handy when the antenna installation must be checked.
If this receiver operates on batteries it is possible to take it up to the
roof and orient the antenna for best reception on all channels in

that manner. Should the installation suffer from unavoidable reception problems, such as ghosts or interference on a particular channel, the portable TV set will demonstrate to the customer that the defect is not in the new color set which he has purchased but is due to his location.

The multimeter serves many purposes. Should any defect crop up in the color TV receiver, the multimeter can be used to determine if it is something simple, such as a fuse or tube, or if the receiver must be returned to the shop for troubleshooting. The multimeter is also very handy in measuring the AC line voltage to make sure that its amplitude is sufficient to operate the color TV set properly. If the line voltage is very low, this can be pointed out to the customer. A step-up transformer can be used or else some changes in the home wiring can be made. When a color bar generator is brought along, it helps to test the performance of the color TV receiver and, if color TV pictures are not available during the installation time, the color bar pattern helps to demonstrate the operation of the set and each of its controls to the customer.

Listed below are the important steps which make up a typical color TV installation. Following a procedure of this type will give the serviceman a set routine and avoid fumbling or brilliant afterthoughts from embarrassing him in front of the customer. There are other routines, following a different sequence, and other installation practices which cut some of the corners, and each individual will have some variation of his own. The procedure below is intended merely as a guide to help the reader develop his own installation technique.

Step 1

Check the set in the shop. No reputable service organization will overlook this step since occasional shipping damage, production errors, and so forth, are unavoidable and the discovery of such a defect in the customer's home provides not only a poor impression, but also costs money in the form of a repeat trip. The extent of the check-up in the shop varies according to the setmaker's reliability, receiver performance, and the available time in the shop. In general, the minimum test includes both monochrome and color reception on all available channels. Adjustments of all front panel and secondary controls is also mandatory. Many shops open the cabinet and check for loose or broken parts which may barely have survived shipment. Making sure that all tubes are firmly in their sockets avoids the possibility of losing a tube between the shop and the customer's home. A careful inspection of the cabinet for scratches and other damage before delivering the receiver is also indicated.

Step 2

Select the right antenna. Usually the monochrome receivers in the neighborhood are a good indication of the type of antenna needed for the new color receiver. Unless the reception conditions are very poor, any antenna that has a relatively flat response at each of the channels received will be satisfactory. Where a black and white receiver was installed previously, the color set may need a new antenna.

This is particularly true whenever the installation is old, the antenna corroded or when the transmission line is brittle and no longer in good condition. It often happens that the customer has become accustomed to some minor ghosts and interference on monochrome reception but will object to them when he sees them in color. In these cases the portable monochrome receiver is very handy because it demonstrates to the set owner that the ghosts are not due to his old monochrome set or due to the new color set. The thing to watch out for is a fringe area installation which uses a sharply tuned antenna such as a Yagi or other narrow bandwidth type. In this instance a new antenna will be required. It is often best, however, to try the existing antenna first and then demonstrate, if necessary, to the customer, that the color set needs a new one. (For a detailed procedure on antenna installation, see *Elements of Television Servicing*, Marcus and Gendler, Prentice-Hall.)

In a completely new TV location many service organizations use a temporary, mobile type antenna first until the best antenna type and location is determined. With the addition of the bandwidth requirement, all of the criteria of a good antenna lead-in line installation apply in color as well as in black and white.

Step 3

Location in the home is a very important factor. A location in which no direct light falls on the screen should be selected. Other factors that will determine the location in the home are maximum seating capacity in front of the set, arrangement of the furniture, and a general harmonizing effect of the entire room. Much could be written about the many possibilities of matching furniture, styling, color harmony, and special effects. The most frequently used arrangement will locate the color TV set either in the spot formerly occupied by a mono-chrome set or simply in a convenient corner.

One successful service organization we know has a little prepared speech which the installing technician delivers to the customer. Its essence is this: "You now have the latest and most wonderful miracle of electronics in your home. You will want to show it to all guests, watch it yourself, and make it a part of the home. Therefore you should locate it prominently in the living room, playroom, or other convenient area. If this means rearranging your furniture, we will gladly help you move the heavy pieces. Keep direct lights away from the screen. Possibly a lamp can be put on top of or behind your color set."

The installation of the antenna lead-in and the power cord arrange-ments are usually secondary considerations to the overall effect of the new set in the home.

Step 4

Test the set on all available channels in the home. Be sure to try it on color as well as on monochrome. If no color program is available, use the color bar generator to demonstrate the correct color per-formance of the set.

Step 5

After making sure that all channels and all controls operate correctly, the customer is instructed in the use of his set. This last step is one of the most important of all five steps. The average customer has little understanding of colorimetry and electronics and the complexities of a simple radio are already far beyond most set owners. For this reason it is essential that all possible care be given to instruct the customer as thoroughly as possible in the use of each control. If some difficulty is encountered do not hesitate to write down exactly which control does what. Most of the TV manufacturers provide simple instruction sheets or tags for the set owner but often these are not sufficiently clear. In any event, it is a good policy to try and instruct at least two people in the household, preferably including some of the younger members. Children are quite technically minded and usually have a better memory than their elders.

At the end of this chapter further details are given regarding the instruction of the customer. More than 60 per cent of all callbacks concerning new TV installations are directly due to customer misunderstanding and only 40 per cent on account of genuine technical defects.

Antenna Installation Problems

As pointed out in Chapter 8, the major requirement for color TV —concerning the antenna and transmission line—is bandwidth. In addition to bandwidth, freedom from reflections and fading, as well as the presence of strong signals, is also necessary for a good color picture. For this reason most of the indoor or built-in antenna types are of limited use for color TV. Only in exceptionally good signal areas will a built-in antenna deliver a really satisfactory signal. The method of tuning indoor antennas for each channel by adjusting the length or the phasing of the elements is not desirable for color because mistuning will result in a non-linear frequency response which in turn will introduce phase errors into the chroma signal.

Outdoor antennas are most frequently used for color TV installations. Whenever more than one station is received or a problem of antenna orientation exists, the color telecasting station will invariably prove to be the most critical as far as reception goes. For this reason most technicians orient the antenna for best color reception and allow the black and white picture to deteriorate slightly. Where an antenna rotator is used, the service man should carefully try it out for all channels and then leave a written record for the customer as to which direction is best for which station. If left to his own devices the customer may select the wrong direction for a particular channel. Although satisfactory for monochrome reception, this direction may not be good enough for color pictures.

Similar limitations are found in some of the multiple antenna installations used in apartment houses and hotels. Wherever insufficient bandwidth is encountered, the only correct remedy is to improve the frequency response so that the entire 6mHz channel is reproduced. In some of the apartment house installations phase shift and reflections due to poor impedance match in the distribution network will ruin color reception. A detailed discussion on color ghosts or reflections is presented in Chapter 19.

Unpacking the Set

This chore is usually performed at the shop, often before the set is sold. Although it seems a simple enough thing to do, improper uncrating of color TV sets accounts for considerable breakage and quite a few adjustment problems. The set manufacturer often provides detailed instructions for unpacking to avoid damage and to make the job as simple as possible. Be sure to follow such instructions whenever available. For the instance in which no detailed crating data is supplied, the general procedure outlined below can be used.

In some instances the receiver and the picture tube assembly are shipped separately. Unpack the receiver cabinet first. Observe whether there is some indication on the carton as to which is the top. Place the carton right side up before even starting to break the sealing tape. Then open the top only after making sure that there are no bolts holding the cabinet from the bottom. The general packing methods used for monochrome TV sets are also used for their color counterparts. Most console cabinets, for instance, are mounted on a wooden shipping pallet while table models are simply packed in a heavy carton with spacers or lining as needed.

After the main cabinet assembly is unpacked, it should be readied for the color picture tube and its components. The picture tube generally mounts in an arrangement similar to that shown in Figure 12.1, the rear view of the Motorola chassis TS-918A. Note that plugs are supplied which connect the various components to their respective signals. Be sure all of these plugs are properly connected before turning power on. Connecting the HV to the picture tube without plugging in the deflection yoke, for instance, could damage the picture tube.

Performance Check

After the entire receiver has been unpacked, all components mounted and all connections completed, an initial performance check will indicate whether the set operates satisfactorily or whether considerable adjustment, alignment, and so forth are required. In this discussion only the case of minor adjustments of the various controls is covered; detailed procedures for the major alignment jobs are the subject of the following chapters.

Using a darkened room if possible, connect the antenna and power cords to the receiver and turn it on. After a few moments' warm-up, tune in the strongest local station and check reception. Although the exact sequence of adjustment may vary, the major steps for setting up good pictures are outlined below as applicable to any color set. Manufacturers' instructions should be followed closely whenever available.

1. Turn chroma control down and use only black and white picture at first. Set the AGC, contrast and fine tuning control for best picture on all channels received.

Figure 12.1—REAR VIEW OF COLOR SET

Motorola

168

2. Check the action of the horizontal and vertical hold control and adjust width and height for best linearity. These adjustments are the same as on a monochrome receiver and are usually located behind a panel as secondary controls.

3. Vary the brightness and contrast control over part of their range to see if picture size or coloring (on black and white) change considerably. If the brightness setting affects picture size, the HV regulator needs adjustment.

4. Turn contrast down and observe raster only. Adjust brightness and the picture tube screen or background controls until the raster appears a dull neutral gray without any color. If color is present in a portion of the raster the color purity needs adjustment. Degaussing may be required.

5. Set brightness control and background controls for good white raster, then dim raster to gray again. With proper color balance the raster should show no coloring at either the dull gray or the white setting.

6. Turn contrast up for good monochrome picture. Check for convergence by looking at the edges of dark objects. If three or two colors appear at these points, convergence adjustment is required.

7. After adjusting for best monochrome picture, turn chroma control up to obtain a colored picture. If a color telecast is not available, tune to a station which transmits a color stripe, at the top of the monochrome picture. Set the vertical synch control so that the stripe is visible. This stripe gives some indication of color performance, even though it is not a complete test. The color bar generator can be used to check color reception.

8. Adjust the color tint or hue control for best colors. This adjustment is somewhat tricky as the correct colors are not always obvious. The best standards are known colors in a telecast. For example flesh tones, sky colors, trees and other natural objects should be used as a guide rather than clothes or interior decorations. It is very difficult to decide whether a shirt is pink, white, or light blue, on the color TV screen, but a green sky or a magenta tree will appear as obviously wrong. The test pattern of known colors should be used, provided by a color bar generator.

If the adjustments outlined above result in satisfactory color and monochrome pictures, the set is almost ready for delivery. Before putting the cabinet into the delivery truck, it is a good idea to check it once more for scratches, missing knobs, screws, clips, loose connectors, brackets, and similar oversights.

Installation in the Home

In general this part of the operation is very similar to a monochrome installation. Since the reader will be familiar with antenna and lead-in mounting, bringing the line through the window or wall into the room, and so forth, these steps will not be enumerated here. Locating the set in the home has also been discussed in this chapter. The tryout of the set at the shop frequently insures trouble-free operation at the first check in the home, but a thorough test for monochrome and color on all available channels should be performed in the home.

One of the peculiar aspects of color television is the purely subjective character of colors themselves. Adjust a color set for best appearing colors in daylight in neutral color surroundings and then view the same set by the subdued lighting of a modern living room with the background of some definite decoration scheme. The difference in apparent color on the TV screen will be quite considerable. For this reason it is advisable to retouch some of the controls after the installation is completed and the set works correctly. If this readjustment is performed during a second visit in the evening when most of the family is home, an opportunity presents itself to demonstrate the set and clear up any customer misunderstandings.

Assuming that the linearity, centering, convergence and purity adjustments are satisfactory and good reception is obtained on all monochrome channels, the remaining checks should concern color only. Carefully set the hue or color tint control and touch up each of the color gain controls very lightly until the best apparent color picture on one channel is obtained. Now, if available, try to get color pictures on some other channel. Retouch the color controls and the hue setting for best color pictures on all channels. In some instances a compromise must be made with color fidelity since one or two stations sometimes vary in their encoding matrices and their signals are slightly different when received.

It is good practice to watch the receiver in the home for about 15 to 20 minutes after warm-up to make sure that there is no drift in either color synch or RF oscillator. When the serviceman is finally convinced that the receiver is performing correctly, the last part of the installation procedure, the instruction of the customer, is started.

Instructing the Owner

Since the ultimate aim of any TV service job is to satisfy the customer, the installation man's instructions to the owner are as important as any of the purely technical parts of the installation and adjustment procedure. Listed below are the most important points which the set owner should know. Considerable tact and human relations knowledge are often needed to avoid offending or slighting the customer. Try to remain polite but firm.

1. Explain that there are several sets of controls on the color receiver, meant for different types of adjustment. The only set of controls which the customer should ever touch is the first set, the front panel controls.

2. Make a strong point of the fact that any of the secondary or tertiary controls can be adjusted only by the expert, some of them only when special test instruments are available; point out that customer adjustment of these recessed controls will require a service call. State clearly that such calls cannot be made repeatedly within the service warranty without some extra charge. Also point out that layman's adjustment of these controls may result in some defect requiring repair.

3. Explain and demonstrate the on-off switch. Point out that there is some warm-up time.

4. Show how to tune in a channel. Make a strong point of the correct fine tuning setting by illustrating "off-channel" picture faults. The need for good fine tuning on each channel cannot be overemphasized with a color receiver. Also point out that warm-up time will often require a resetting of the fine tuning control.

5. Demonstrate the volume and tone controls. Usually these present little difficulty to the customer.

6. Turn the chroma control down and show how to set both brightness and contrast control correctly. In monochrome, incorrect setting of these two controls does not wreck the picture as much as in color TV; greater care must therefore be taken to explain the meaning of the brightness and contrast controls to the customer. Let the customer try both controls and give you his idea of a correct setting. Do not start with the chroma control until at least some of the members of the family understand and are able to perform the monochrome adjustment correctly.

7. Turn up the chroma control and explain its function to the customer. Point out the oversaturated color condition and the insufficiently saturated picture.

8. If there are any other front panel controls, explain them as well. In some receivers all of the controls are behind a panel and some secondary controls are also accessible. This allows the customer to make such adjustments as green and blue video gain, green and blue background brightness, and color phasing. Since the average set owner does

not have a sufficient technical background, adjustment of these controls will not only confuse him, but also result in the wrong color pictures. With the owner's consent, these controls are often taped off to avoid even accidental adjustment by the customer.

9.   After the installing technician has completed his explanations several members of the customer's family should be permitted to tune in color and monochrome signals and perform all of the primary adjustments. Point out mistakes and show the correct settings for each control as often as necessary until each adjustment is clearly understood.

10.   Before leaving, it is a good idea to ask if there are any questions. This is not only a polite way of finding out if you were understood, but also may avoid later calls for relatively minor things. Leaving the company's or your own card and your own name completes the average installation.

Much has been witten in this chapter about installation problems which are not of a technical nature. Experience will show that these non-electric aspects of color TV servicing account for at least as many service calls as the purely technical breakdowns. The following chapters deal in detail with the practical TV alignment and troubleshooting problems of color receivers.

13

RF-IF ALIGNMENT

In the previous chapter the procedure for installing a color television receiver in the home was discussed. There it was assumed that the receiver was properly aligned and operated correctly. Unfortunately, this is not always the case. Often the manufacturer's quality control fails and receivers are delivered with poor RF and IF alignment or else damage occurring in shipment requires a repetition of the entire alignment procedure. In some instances defects occur during the normal life of the receiver which make alignment of one or more of the receiver sections necessary. This chapter concerns itself solely with the process of aligning the RF and IF sections. Chapter 8 describes the circuitry and performance requirements for the IF and RF portions of the color receiver; in this chapter the methods and some important points of the alignment itself are discussed.

Tuner Alignment

In modern TV servicing the most frequently used method for aligning broadband amplifiers such as the RF and IF section is the visual method. Few servicemen depend on the point-by-point method with a CW signal generator and VTVM as indicating device. For color TV the latter method is too time-consuming and not sufficiently accurate. The method used and recommended universally employs a sweep frequency generator and an oscilloscope, on which the response curve appears. To specify individual frequencies, a marker generator is used or else the sweep generator contains its own accurate markers.

The sweep generator acts like any oscillator which is rapidly tuned through a certain range of frequencies. The rate of frequency variation is usually 60 Hz and the range of frequencies is any 10 or 15 mHz band which includes the TV channel to be aligned. When the output of the sweep generator is detected and displayed on an oscilloscope, the vertical axis corresponds to RF amplitude and the horizontal axis represents frequency. Thus by passing the sweep generator signal through a bandpass network, the response curve of that network appears on the oscilloscope screen. The identical method of visual alignment is used for monochrome TV work.

Two distinctly different types of alignment are necessary in the tuner. One is the adjustment of the oscillator frequency to obtain the correct IF beat, and the second is to align the bandpass of the RF amplifier for correct frequency response characteristic. In general it is best to align the RF section first and set the oscillator later. The reader will be familiar with the individual channel frequencies from the monochrome alignment and from local channel assignments. For alignment purposes there are three TV bands. Channels 2 to 6 occupy the lowest band, from 54 to 88 mHz. Channels 7 to 13 range from 176 to 216 mHz and channels 14 through 83 are in the UHF band from 470 to 890 mHz.

A typical RF bandpass response is shown in Figure 13.1. Note that the portion between the video and sound carrier is practically flat and the overall bandwidth is at least 6 mHz. This is the basic requirement of any tuner for color reception. In VHF tuners there is at least one stage of the RF amplification. Figure 13.2 shows a typical RF stage with the various tuning elements. A large portion of all TV receivers use a turret

Figure 13.1—RF BANDPASS

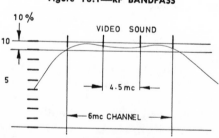

tuner which switches in different coils for each channel and has master adjustments as well, which affect all channels. In a switch type tuner, the coils for all channels are in series as in Figure 13.2 and alignment should start with channel 2. In monochrome receivers it is usually possible to achieve a satisfactory compromise between different channels by only adjusting the master trimmers. For good color performance it may be necessary to touch-up the individual coils in a particular segment.

Figure 13.2—RF STAGE IN VHF TUNER

Sony

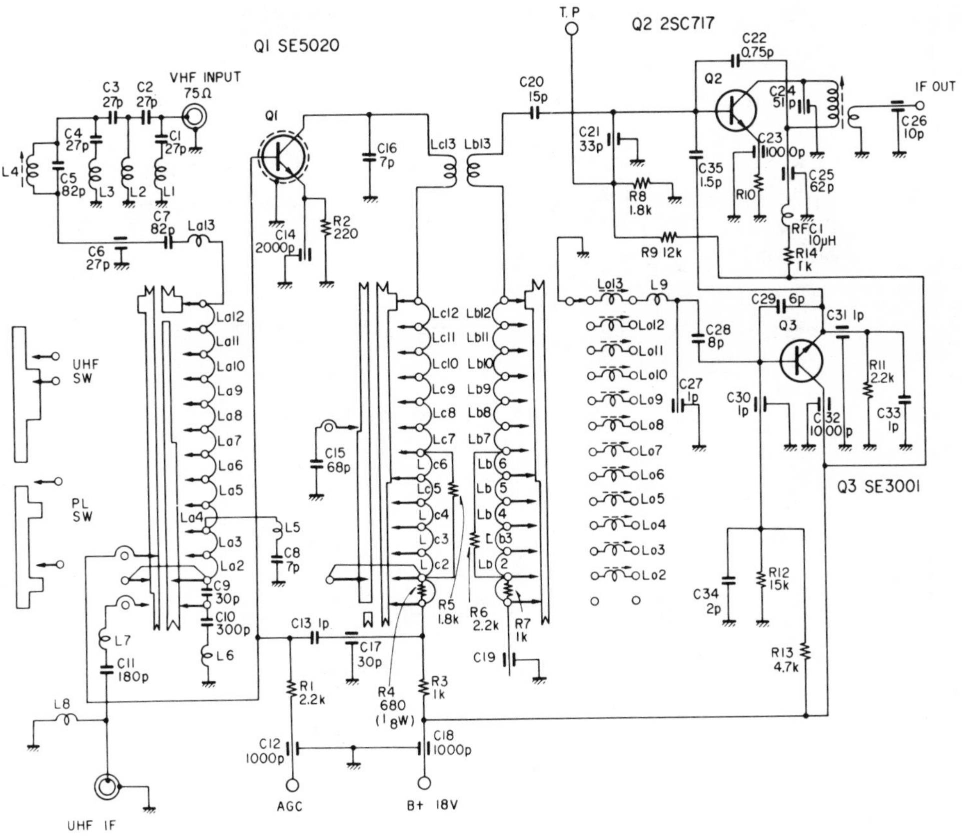

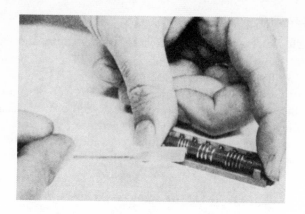

Figure 13.3—RF COILS IN TUNER SEGMENT

Figure 13.3 shows a typical set of RF coils mounted in a turret tuner segment and indicates how adjustments are made by moving turns closer or spreading them. Other tuners, such as the Sarkes-Tarzian or Oak Mfg. Co. switch types, also allow trimming up individual coils. The drawback of the turret is the fact that the adjustment is usually made with the segment removed from the turret so that its effect cannot be immediately observed.

A typical RF tuner alignment uses the block diagram shown in Figure 13.4. Note that the vertical oscilloscope terminals go to the test

Figure 13.4—BLOCK DIAGRAM FOR TUNER ALIGNMENT

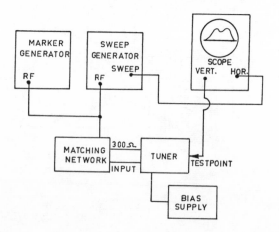

point on the tuner. Since the mixer is a non-linear element it acts like
a detector, and, by using a small isolating resistor, the detected 60 Hz
signal for the scope can be taken from there. A shielded lead should
be used to avoid pickup and to remove some of the RF signal.

It is quite rare that the frequency response curve of the RF stages
of any tuner will be as flat and neat as the one shown in Figure 13.1.
More likely the response may appear like the one in Figure 13.5 which,

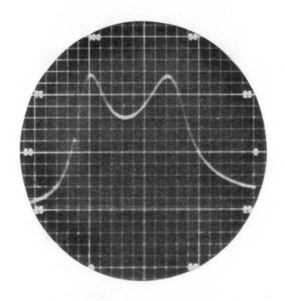

Figure 13.5—RF BANDPASS NOT FLAT ENOUGH, DEEP VALLEY

while satisfactory for monochrome reception, is not flat enough for
color TV. Careful adjustment of the master trimmers and, if necessary, individual channel coils can usually achieve sufficient flatness.
Remember that a deep valley is often caused by too much coupling
between two coils. For the response curve in Figure 13.5 it would
therefore be advisable to separate the RF input or output coils slightly
and reduce the coupling. Another objectionable characteristic is shown
in Figure 13.6, which illustrates the appearance of a response curve
when no baseline blanking is used. The major objection to Figure 13.6
is the fact that one peak is considerably higher than the other. Referring back to Figure 13.1, note the two dotted horizontal lines enclosing

the flat portion of the response curve. These two lines indicate the tolerance for valley, tilt, and other variations in response curve. By limiting any such deviation to less than 10 per cent of the total amplitude the limits of phase shift and time delay in the color section of the receiver will be maintained. To make sure the 10 per cent limit is not exceeded, adjust the oscilloscope vertical gain so that the response curve is about 20 divisions high. Now observe if the curve between the sound and video carrier does not vary for more than two divisions.

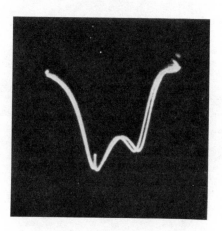

Figure 13.6—RF BANDPASS, NO BASELINE

The illustration in Figure 13.5 exceeds the limit, as does Figure 13.6.

While the VHF tuner alignment should present no problems to the technician familiar with monochrome TV sets some precautions and suggestions are in order. One important factor is the proper impedance match of the output of the sweep generator and the tuner input. Practically all VHF-TV tuners use a 300 ohm input circuit, but some of the sweep generators use coaxial output leads with 75 or 50 ohm impedance. To match this to the tuner there is usually a special resistor-type matching network which is part of the output cable. Where such a network is not supplied with the generator it is not too difficult to construct a simple resistive 75 to 300 ohm pad as is shown in Figure 13.7. Another thing to watch for is the generator output amplitude setting. When this is excessive the generator signal will overload the amplifier and give the appearance of a very flat response

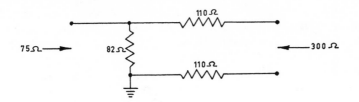

Figure 13.7—75 TO 300 OHM RESISTIVE PAD

curve. Be sure to vary the amplitude and set it at that level at which the response curve is not flattened out.

The amount of bias introduced by the AGC circuit controls the amplitude and, in some tuners, the response characteristic to some extent. For the latter case the best remedy is to check what the AGC voltage is for each station and then align the tuner with a bias corresponding to the respective AGC voltages.

Make sure there is no RF oscillation due to the RF amplifier by reducing the generator output to zero and checking both the scope and the TV screen for any signals. Cascode RF stages are especially susceptible to oscillation when some of the coils have been changed without compensating the neutralizing circuit. Oscillation of a cascode stage can usually be corrected by either adding or reducing capacity from the neutralizing network.

Alignment of the RF stage in a UHF tuner often presents more problems. The major difficulty is to get good tracking over the entire UHF band. Since most localities rarely receive more than two or three UHF stations it is usually possible to get fairly good response curves at the few selected channels if the rest of the band is neglected. For color TV receivers this procedure is often required in order to get satisfactory reception on UHF.

Needless to say that, although the connections are the same as for the block diagram of Figure 13.4, the frequency range of the generator must be different for UHF alignment. Most UHF tuners do not employ an RF amplifier, but simply have a tuned input filter, using either two or three resonant networks.

Most modern UHF tuners employ a transistor as local oscillator and a crystal diode as a mixer. The output is at the IF frequency and goes to the VHF tuner which acts as first IF stage when set to "UHF." If the oscilloscope is connected to the same test point as for VHF alignment, this means that the response at the 40 mHz IF frequency will include the response of the VHF tuner. If the oscilloscope is connected to the output of the UHF mixer diode itself, it is usually necessary to provide an RF probe to demodulate the IF signal at that point. The amplitude of the signal at the crystal diode output will be considerably less than at the VHF tuner test point. In either case it is important that the output amplitude of the UHF sweep generator be adjusted to avoid overloading. Too much sweep generator signal will tend to saturate the UHF mixer diode and this would present a false picture of the frequency response. If the RF input circuit is badly misaligned, very little signal will pass. If a weak response curve is obtained, the UHF tuning control should be reset to make sure that gross misalignment is not the case.

Important as the impedance matching between generator and tuner is in VHF, it is still more critical for UHF work. Since the signal strength requirements are greater for UHF, resistive matching networks are not always used and many sweep generators are equipped with external "Baluns." This is an impedance transformer, changing coaxial 50 or 75 ohm levels to a balanced 300 ohm system. Be sure to use the right balun combination since a mismatch between 50 and 75 ohms is sufficient to give erroneous results.

Bias is generally not a problem in UHF tuner work since no AGC signal is supplied to the UHF section.

UHF response curves are generally much broader and usually also flatter than their VHF cousins. It is important that more than just the video and sound IF carrier be on the flat portion of the UHF response curve since the customer usually does not tune in the UHF channel quite as simply and accurately as the VHF. Oscillator drift at UHF is also more of a factor than at VHF. In general, it is a sound rule to try for at least 10 mc flat region with the desired UHF channel centered in this band as shown in Figure 13.8.

Figure 13.8—UHF BANDPASS—10 mHz FLATNESS

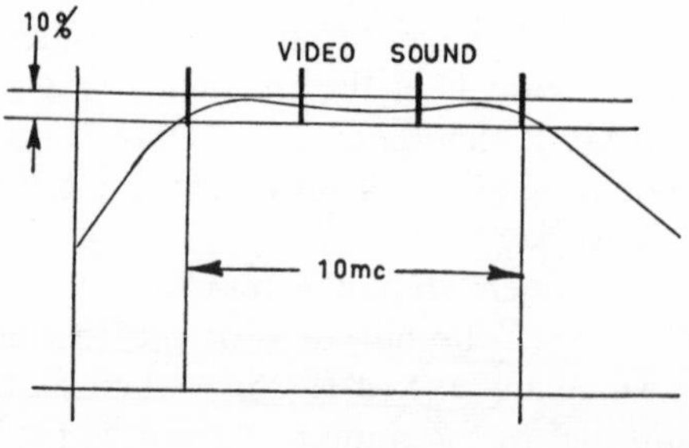

Oscillator Alignment

The local oscillator in both the VHF and UHF tuner should be aligned accurately after the IF alignment is completed. Since this part of Chapter 13 already deals with the tuner, the adjustment of the oscillator is described here with the assumption that either the IF section has just been aligned or else is known to be operating perfectly. The correct sequence for aligning a color TV set is to start at the RF stage, then align the IF section, including the sound IF, and finally, check the oscillator adjustment.

The oscillator alignment for turret type VHF tuners is relatively simple since each coil board contains a small tuning slug which can be adjusted from the front of the TV set. As a matter of fact, this alignment is the same as for monochrome sets and usually can be made without removing the chassis from the cabinet. Switch type VHF tuners usually have a master oscillator trimmer as well as separate fine trimming adjustments for each channel. Where automatic fine tuning (AFT) is provided, be sure to disable the AFT circuit before aligning the oscillator.

For either type of VHF tuner, simply switch to the desired channel and adjust the oscillator for best picture on that channel. To make sure this adjustment is correct, allow at least 10 minutes warm-up time and then set the fine tuning control to a mid-position. A close check on oscillator adjustment can be made by using a signal generator tuned accurately to the sound RF carrier. When a VTVM is connected at the video detector output, align the oscillator until the sound RF carrier signal falls in a sharp steep dip. In other words, the VTVM will show a sharp zero. To verify this action, add the RF sweep generator and observe the overall response curve of the receiver on the oscilloscope. When the oscillator is set correctly, the fine tuning control will be able to place the sound RF signal right into the sound IF dip. Another method uses a station signal with VTVM connected at the intercarrier limiter grid. Adjust the oscillator for a peak indication on the VTVM. One of the most important features of the oscillator alignment is the range and drift compensation for each channel. The fine tuning control should be able to provide this compensation, but since the oscillator adjustment for each channel governs the center frequency of the fine tuning control, this center frequency setting should be as accurate as possible.

On UHF tuners the oscillator adjustment depends greatly on the type of tuning mechanism used. For continuously tuned UHF networks, usually only a master trimmer control is provided. In this instance the alignment process involves first tuning for correct oscillator frequency and then checking the RF bandpass to provide for optimum flatness. Bandpass characteristics as well as the range of the fine tuning control or the vernier tuning knob are usually broad enough to permit correct tuning.

IF Alignment

One of the most important new aspects of television as compared to radio was the introduction of the broadband IF section. Basically the same IF section is used on color TV receivers. The major differences in the new IF sections were described in detail in Chapter 8 and involve mostly tighter tolerances and a few new trap circuits. Alignment procedures are also basically the same, but greater precision and care is required. A typical IF circuit was shown in Figure 8.11 and 8.12; the desired response curve is indicated in Figure 13.10. In order to obtain these results, the visual alignment method, with crystal controlled markers, is almost invariably used.

Figure 13.9 shows a typical set-up of the instruments and cable connections required. Note that a variable bias supply is included. This

Figure 13.9—BLOCK DIAGRAM, IF ALIGNMENT

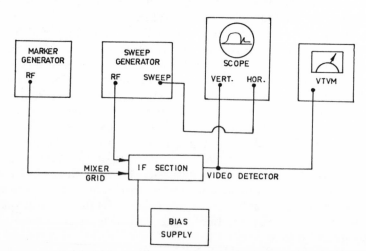

may consist of four flashlight cells in series, providing a total of 6 volts with a 100,000 ohm potentiometer shunted across it to allow varying the bias to any desired value. The bias supply is connected across the AGC bus and should be metered for accurate checks. The typical alignment procedure shown below does not supersede the alignment data supplied by the set manufacturer, but rather is intended to supplement this data. Frequencies given here apply for the standard 41 mHz IF system used in the majority of color receivers.

Alignment can be conveniently divided into two parts: the trap setting and the overall alignment. For the first part the sweep generator is not needed and an AM signal generator or crystal controlled marker is used with the VTVM or scope.

Trap alignment:

1. Set the bias supply for about 3 volts.

2. Tune the generator to exactly 41.25 mHz. If the VTVM is used as indicator, connect it to the video detector load. If the scope serves as indicator, the signal generator or marker generator should have some AM component such as 400 Hz or 1000 Hz.

3. Connect the generator to the grid of the mixer and adjust each of the sound IF traps, in turn, for minimum indication. When it seems that the minimum point is indistinguishable due to insensitivity, reduce the bias to 1.5 volts or even to zero until the indicator (VTVM or scope) shows only the absolute minimum setting. After setting each trap for minimum, go over them again for a slight touch up.

4. Set the generator to 47.25 mHz and tune those traps in the same manner. In some receivers one of the 41.25 mHz networks and the 47.25 mHz traps are connected in series and adjustment of one may have some effect on the other. This requires retouching.

5. If a trap for the adjacent channel video signal is used, this is usually tuned to 38 or 39 mHz and the generator should be set accordingly.

Occasionally there is some difficulty in aligning one of the high Q traps for the sound IF which tends to start oscillations or else appears not to tune correctly. Isolate the particular stage by applying the generator to the preceding grid and connecting a crystal diode probe to the grid following that stage. This method indicates whether a particular component is defective, or whether the entire IF section is regenerative.

In some early color TV sets the bridge-T traps used for the sound IF also used a potentiometer to adjust the series coil resistance and thus balance out the bridge. In such a case the potentiometer is set after the coil is tuned for optimum sound rejection, and the coil may require slight retouching.

Bandpass Alignment

1. Set the sweep generator for maximum sweep width and adjust the bias control for about 3 volts.

2. Set the sweep generator output amplitude to zero and then bring it up gradually until the response curve seems to overload. Then reduce the output slightly.

3. Set one marker to 45.75 mHz and the other marker to 42.2 mHz. If only one is available, mark the spot where the 45.75 mHz point falls on the screen with grease pencil, and set the marker to 42.2 mHz.

4. Following the individual coil data given by the manufacturer, set each stage to its approximate frequency. To do this it may be necessary to move the marker to the various alignment points. If it appears that some of the coils are considerably misaligned, adjustment of individual stages may be necessary.

For a stage-by-stage alignment, connect the sweep generator to the input of the last IF stage and adjust the last IF coil for its correct frequency. Next connect the generator to the preceding stage and repeat for that circuit. In this manner each stage is individually aligned for its correct frequency and then the overall IF response needs only minor adjustments.

In color TV it is important that the phase delay as well as the frequency response be correct. This means that the technician cannot simply adjust all stages until a fairly decent IF response is obtained; each stage must be tuned according to the original design.

The overall IF response curve may appear like the one shown in Figure 13.10 when all touch-up adjustments are made. Now comes the test of correct alignment. Vary the bias from 6 to 1 volts and see, after adjusting the sweep generator output and the oscilloscope gain, whether the frequency points of 42.2 mHz and 45.75 mHz still fall on the flat and 50 per cent portion of the curve respectively. Check the height of any dip between peaks and be sure it is less than 10 per cent. Also check the location of the various dips at the sound and adjacent sound traps.

If the alignment is correct, it should be possible to vary the bias from 1 to at least 5 volts and still get essentially the same response curve. At higher bias values, more IF signal will be required while at lower values the vertical scope gain may have to be increased. One criterion of proper alignment is the constancy of the various alignment points with variation in bias. If it does not appear possible to get close tracking in this respect, make sure that at least the sound IF dip and the color subcarrier side band position on the curve is fairly constant.

The IF sections of color TV receivers are basically the same as for monochrome and all of the defects inherent in them can occur. Lack of gain, regeneration and oscillation are common IF troubles and both source and cure of these defects are the same as in monochrome sets. Since practically all color TV sets use the 41 mHz IF band rather than the old 21 mHz region, the higher frequency makes the IF section more susceptible to regeneration and radiation troubles. The usual precautions of shielding, decoupling of filament and B-plus leads and good grounding apply. In the alignment procedure be sure that none of the leads are "hot" and that good connections exist between the set and all associated instruments.

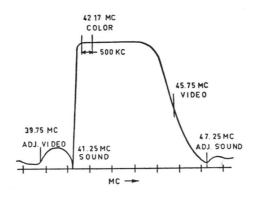

Figure 13.10—CORRECT IF RESPONSE CURVE

Sound IF Alignment

In practically all color TV receivers the sound carrier is amplified along with the video signals and the second sound IF is 4.5 mHz as in the conventional intercarrier system. The separation of sound and picture signals is somewhat tricky in color receivers since the color subcarrier is only 900 kHz away and any beat between the two signals would appear quite objectionable on the screen. To avoid such beats, the sound IF, usually 41.25 mHz, is attenuated at least 50 db with respect to the color sub-carrier in the IF section and at the detector the 4.5 mHz sound IF is removed by a high Q circuit. Alignment of the sound take-off coil at the detector presents something of a problem when no station signal is available.

One method of aligning the entire sound IF section is to connect two signal generators, set at exactly 41.25 mHz and 45.75 mHz respectively, at the grid of the mixer stage. This will result in a 4.5 mHz beat at the video detector, and it is this beat signal for which the sound IF section is tuned. Connect the VTVM to the input of the limiter stage or else to the AVC point on the ratio detector. Then simply tune the sound take-off coil at the video detector for maximum signal. If the sound take-off coil is part of a 4.5 mHz trap circuit, adjust the trap also for maximum VTVM indication. Align each of the sound IF coils and transformers for maximum bias at the limiter. Without touching anything, next adjust the ratio detector primary for maximum signal at the center and align the secondary for zero voltage at the audio output terminal. To check the ratio detector alignment, vary the signal generator tuned to 45.75 mHz slightly in frequency. The VTVM should show positive and negative voltages at the audio output terminal as the generator frequency is shifted.

With a precise 4.5 mHz signal available from the generator, it is a simple matter to tune the primary and secondary of the FM detector transformer or, in the case of the beam deflection type detector tube, the screen and control grid tuned circuits. The transformer primary or the control grid coil is tuned for maximum output and the secondary or screen grid coil for zero, as in any FM receiver. The 4.5 mHz signal generator method is preferred for the alignment of any type of inter-carrier sound section, whether it is a color or monochrome TV receiver.

When a station signal is available either in color or monochrome, it can be used for the sound IF alignment. Simply connect the VTVM to the limiter input or AVC bus and adjust the sound take-off coil and the subsequent tuned networks for maximum voltage. Alignment of the ratio detector with a station signal is somewhat tricky since the exact zero point cannot be readily located. By carefully tuning the secondary of the transformer to the best sound reproduction, a close approximation to the perfect alignment can be achieved. The ratio detector transformer primary is generally tuned for loudest as well as clearest sound, but the use of the 4.5 mHz signal from the generator is the preferred method.

In most color TV receivers there are one or more 4.5 mHz traps in the chroma section or the brightness amplifier. The alignment of these networks is taken up in the following chapter which deals with these two sections.

14

COLOR DECODER
ADJUSTMENT

This chapter deals with the color TV receiver sections handling the video signals that ultimately produce the colored image on the screen. In other words, we shall concern ourselves with the brightness amplifier channel, the chroma amplifier, and the color demodulator and matrixing networks. The operation of these circuits was described in detail in Chapter 9, and here only the alignment and adjustment procedure will be taken up.

In many instances the alignment of the decoder section will be performed as part of the overall adjustment procedure after major repairs. In this case the RF and IF alignment is performed first and the synchronizing adjustments described in Chapter 15 should be set at least roughly before the decoder is adjusted. Most often, however, the RF and IF sections may be aligned correctly and only the decoder section will require tuning up for best color pictures. In either event, it is assumed that the RF and IF sections are correctly aligned before the adjustment of the decoder is made.

Although the decoder circuits vary in different color receiver models, the four basic areas of adjustment are as follows:

1. The brightness channel
2. The chroma amplifier

3. The color demodulator
4. The matrixing and video amplifier section.

In some receivers the matrixing and video amplifier section are so designed that only a minimum of adjustment is possible and in the majority of sets there is little alignment necessary in the brightness channel. For troubleshooting purposes the method of alignment should be understood so that certain defects which cause misalignment can be located and repaired.

Equipment Needed

For efficient adjustment of the color decoder section, the following test equipment should be available:

Oscilloscope. This scope should have a vertical input frequency response to 4.5 mHz if at all possible. Vertical input sensitivity through the compensated amplifiers should be in the order of 10 millivolts per inch or better. It is possible to use a regular low frequency oscilloscope for some of the tests but this does not permit some of the phase and signal tracing techniques.

Signal generator. This should be an accurate generator, preferably with crystal check points. Frequency range should be from 10 kHz to 5 mHz and output amplitude should be adjustable accurately.

Video sweep generator. Some of the RF/IF sweep generators also have a setting for video sweep range. If the sweep output is from at least 100 kHz to 5 mHz and is reasonably flat, the generator can be used.

Detector. A crystal probe with sufficient filtering for the detection of the video sweep signal is required.

Color bar generator. Any one of the variety of available color bar pattern generators can be used. The major specification is that color signals of known colors be available.

Brightness Channel Adjustment

In earlier chapters we have pointed out that the brightness channel is really nothing more than the video amplifier of a monochrome TV receiver with a few special features. One such feature is the delay network which delays the Y or brightness signal just long enough to make up for the time it takes the color signals to pass through the color section. The second feature in some brightness channels is the presence of one or more traps to remove the intercarrier sound signal. These traps are tuned to 4.5 mHz and produce a sharp dip on the video frequency response. These traps are usually the only adjustable component in the brightness amplifier. The contrast control, which determines the gain of the Y signal, is set for individual station signals and is usually one of the operating controls located on the front panel.

In general the only reason for aligning the Y amplifier section appears when a defect in that circuit is suspected. Such a defect might be apparent in poor high frequency response on monochrome and color pictures, incorrect overlap of the color and Y signals, or smearing or ringing where fine picture detail appears. All of these troubles could be due to defective components in the Y channel and these will show up in any video alignment.

To align the Y amplifier section, connect the video sweep generator, marker signal generator, detector probe, and oscilloscope as shown in the block diagram of Figure 14.1. For this alignment the oscilloscope need not have a wide frequency response since the output of the detector will only be 60 Hz. The frequency response of the Y amplifier should be reasonably flat—to about 4.2 mHz—with a notch at 4.5 mHz.

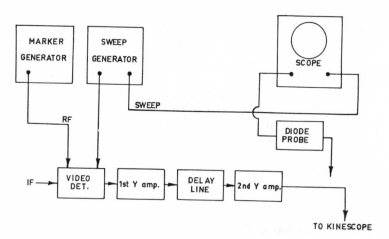

Figure 14.1—BLOCK DIAGRAM Y CHANNEL ALIGNMENT

This response curve may appear as shown in Figure 14.2, where the high spot at the left represents the low frequency end. In many generators the sweep mechanism for video ranges produces a double band due to the beat method used. Be sure to use only one half of the resulting picture for the video frequency response check.

Coupling the signal generator to the sweep generator output can be done in several different ways. The simplest is to connect one side of the output lead to ground and then connect the hot side through a .005 mfd capacitor to the hot side of the sweep generator cable. Then adjust the output amplitude of the marker generator until a pip of proper size appears on the response curve.

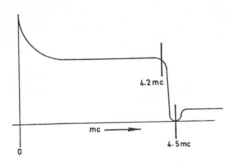

Figure 14.2—Y CHANNEL RESPONSE

To align only the 4.5 mHz trap, the sweep generator is not needed and only the signal generator, the detector, and a VTVM are required. Simply tune the trap for minimum VTVM reading when the signal generator is tuned to exactly 4.5 mHz.

With the video sweep connected and the actual frequency response visible on the oscilloscope, vary the contrast control to make sure that it does not materially affect the response curve.

The delay line operation was described in Chapter 11; it was pointed out there that basically two types of defects can occur. If there is a short in the shunt capacity, or an open in the series inductance, no signal will pass and this defect can be located by simple ohmmeter checks. If there are shorted turns in the inductance or open capacitors, the result will be a loss in some of the delay. At the same time, the flat frequency response of the delay line will also be impaired and this will become apparent when the video frequency response curve is checked. Other defects such as leaky coupling capacitors, open or shorted video peaking coils, and the like, will all result in a deterioration of the frequency response. In this respect the circuit behaves just like the video amplifier of a monochrome receiver. One more consideration of the Y channel is the overall gain. This can be checked by simply feeding a signal, either from the generator or from a TV station, through the video detector and following its amplification with the oscilloscope through each stage. The average gain per stage should be at least 10 times or more, depending on the contrast control setting.

Chroma Channel Adjustment

The detailed circuitry of the chroma bandpass amplifier was described in Chapter 9. Ordinarily this section does not require adjustment, but if a major component has been removed, or if some defect in that section is suspected it may have to be aligned. This alignment simply consists of adjusting the two or three tuned circuits which provide the 3.58 mHz bandpass response and, where used, the 4.5 mHz intercarrier sound traps. When a sweep generator, marker and oscilloscope are available, the actual response curve of the bandpass amplifiers can be displayed. When only a signal generator is available, the VTVM can be used in place of the oscilloscope and the bandpass response can either be plotted from taking measurements at the center and the two extreme points, or else an approximate idea can be obtained by observing the meter as the signal generator is tuned over to the 3 to 4 mHz band. For accurate alignment and really professional service work the sweep generator and oscilloscope method is preferred.

Set the sweep generator for video range and connect it together with the marker generator, detector, and oscilloscope as shown in Figure 14.1. The signal is applied at the video detector and the detector probe is connected to the input of either demodulator. If some kind of AGC system, such as ACC (Automatic Color Control), is used for the chroma amplifier, be sure to substitute a variable battery type bias. Color killer circuits will cut off the chroma stages when no color burst is present and for alignment it is necessary to short out or disconnect the killer circuit. Where a manual color-monochrome control is used, be sure to set it to "color" before starting the chroma alignment.

Figure 14.3—TYPICAL CHROMA RESPONSE

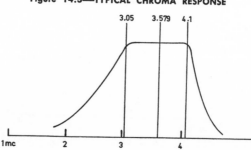

Whenever manufacturer's data is available, it should be followed closely. In the absence of this information the alignment of a typical chroma amplifier can be done in this manner:

1. Align 4.5 mHz traps first for minimum signal output just as for brightness amplifier.

2. Set marker generator to 3.579 mHz and adjust each tuned network for maximum amplitude. Where overcoupled transformers are used in the chroma channel, tune for maximum amplitude first and then touch up for flatness.

3. Vary the chroma control and see if this controls the amplitude without changing the response characteristics.

4. Where an ACC system is used, substitute a battery type variable bias and vary the bias voltage over the normal ACC range to see if the overall response curve remains constant.

In some color receivers the video detector is part of a shielded assembly which contains both IF and video tuned networks. Be sure to adjust only those coils belonging to the video section.

Although alignment of the color synchronizing circuit is covered in the next chapter, it might be convenient to align some of the color synch circuits together with the chroma channel. With the marker generator set for 3.579 mHz and the sweep generator output reduced, the color burst take-off coil can be adjusted for maximum response.

Demodulator Alignment

After the chroma channel has been aligned, the major adjustment at the demodulators is the phasing between the two reference signals. Before this is done, however, it is convenient to set the 3.579 mHz traps in the output of the demodulators. Leaving the sweep generator and marker generator connected as for the chroma channel alignment, merely move the detector probe to the matrixing section, or, in the case of high level demodulators, directly to the kinescope grids. It may be necessary to set the marker generator for maximum output in order to detect sufficient 3.579 mHz signal. When a wide band oscilloscope is available, the sweep generator can be omitted at this point and only the 3.579 mHz signal used. The traps are then adjusted for minimum amplitude on the scope. Be sure to include all the 3.579 mHz traps in this procedure. There is usually one in each demodulator channel, and in the high level demodulation systems there is also a trap in the green difference channel.

The low pass filters at the output of the demodulators are usually of the fixed type permitting no adjustment. If it is desired to check their performance, simply connect the video sweep generator to the input of the demodulators and the detector probe to the matrixing networks. The correct bandpass for the B-Y and G-Y signals is found in Figure 14.4.

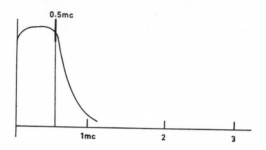

Figure 14.4—COLOR DIFFERENCE VIDEO RESPONSE

In many modern color TV receivers the phase relationships between the two color synch signals applied to the demodulators is determined by fixed components. In the circuits of Figures 9.12 and 9.15, for example, no adjustment is provided for the X and Z phase or the R-Y and B-Y phase. In the beam deflection circuit of Figure 9.20, however, the top and bottom of a special transformer must be adjusted to set the relative phase of the R-Y and B-Y synch signal.

The color bar generator, which is one of the most important tools for color TV servicing, provides the simplest and most effective means of aligning the two synchronizing signals with respect to each other. This alignment should only be performed after the other stages of the color synch section have been properly aligned, a procedure described in Chapter 15.

With the color bar generator and the oscilloscope connected as shown in Figure 14.5 and the tint or hue control set to midpoint, the output of each of the demodulators can be observed on the scope. The adjustment of the color synch going to the R-Y demodulator is tuned until the R-Y demodulator output shows the color bar pattern of Figure 14.6. The zero position of the correct number color bar is the important factor. For the B-Y color synch adjustment the corresponding color bar pattern is shown in Figure 14.6b.

As in the alignment procedure for the bandpass amplifier on the preceding pages, the chroma gain control should be set to a reasonable

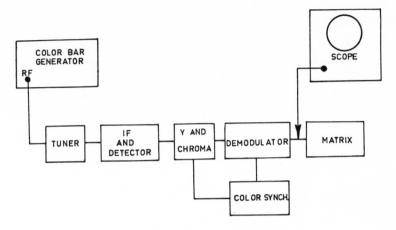

Figure 14.5—COLOR BAR GENERATOR SET-UP

value. It is important to remember that the color bar pattern illustrated in Figure 14.6 may not, at this stage of the alignment, produce the desired color bar pattern on the picture tube screen. The adjustment of the background, CRT screen grid control and others may account for wrong colors appearing on the screen. As long as the color sequence of the gated color bars is as shown in Figure 14.7 and the demodulator output appears as in Figure 14.6, the alignment for the R-Y and B-Y channels can be performed in this manner.

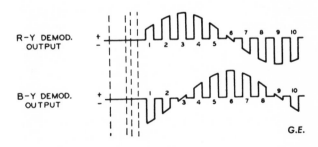

Figure 14.6—MULTIPLE BAR PATTERN IN R-Y AND B-Y CHANNELS

Matrix Adjustment

The matrix circuits of modern color TV receivers ordinarily do not contain any adjustable circuits. In a few models, such as those using the beam deflection demodulators, the 3.58 mHz traps can be considered to be part of the matrix circuit but they are aligned as described above as part of the demodulator. The relative matrixing amplitudes of the red, green and blue color difference signals are fixed in practically all of the color TV receivers produced since 1958. Certain defects in the matrix circuits, however, can give the appearance of a misadjustment of the relative amplitudes of the three color difference signals, and it is for these reasons that a procedure to check the color matrixing functions is outlined below. The color bar generator and the oscilloscope are the only test equipment required for this procedure.

1. Connect the color bar generator to the antenna terminals of the receiver and tune in the respective channel until some kind of color bar pattern is obtained on the screen. Connect the oscilloscope probe to the R-Y control grid of the color picture tube and synchronize the oscilloscope with the color bar generator.

2. With the chroma gain control set to approximately midpoint, adjust the hue or tint control until the third color bar, the red in Figures 14.7 and 14.6a, is maximum amplitude on the oscilloscope screen.

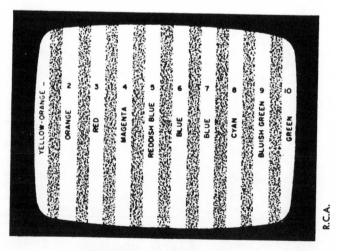

Figure 14.7—COLOR BAR PATTERN

Mark the amplitude on the scope screen with a grease pencil or else note down the number of vertical divisions.

3. Connect the oscilloscope probe to the B-Y control grid of the color picture tube and repeat the procedure of step 2 above for the sixth and seventh bar, the blue. If the red and blue matrixing sections are working correctly, the amplitude of the third bar and of the sixth and seventh bar, respectively, should be approximately the same on the scope.

4. Connect the scope probe to the G-Y control grid of the color picture tube and repeat step 2 above, checking the amplitude of the green or number ten (10) bar as above.

In the above procedures the relative amplitudes of the number 3, 6, 7 and 10 color bars indicate the amount of gain of the red, blue and green color difference signals through the matrix. If it appears that one of these color bars is substantially larger or smaller than the other two, a defect in that particular matrixing section should be suspected. To make sure that it is the matrixing and not the demodulator, the oscilloscope probe can be moved to the output of the demodulator section and the respective color bars can be observed there. If the oscilloscope probe is moved to the input of the demodulator, it will be necessary to use the RF detector probe. At the demodulator input it is possible to observe the entire color bar pattern which should appear of the same amplitude for a complete frame. After isolating the defect to one of the respective circuits, the defective component can be usually located by voltage measurements and ohmmeter tests.

15

SYNCHRONIZATION AND CRT COLOR ADJUSTMENT

This chapter covers all of the remaining adjustments which are ordinarily made during the installation or troubleshooting of a color TV receiver. We shall describe the adjustments of the vertical sweep circuit, the horizontal sweep circuit, the centering, pin cushion, high voltage, and focusing. The alignment of the color synchronizing and color killer section will be discussed in some detail and then the setup of the color purity, color temperature adjustment or the black and white tracking, as it is sometimes called, will also be covered. Probably one of the most difficult and most frequently aligned sections of the color TV receiver controls is the convergence. For this reason the end of this chapter will be devoted in some detail to the various convergence alignments that are required and a typical sequence of convergence adjustments is included.

It is not practical to reprint here the detailed instructions which are contained in the service manuals provided by the various receiver manufacturers. In this chapter, therefore, we shall only cover the general procedures, without reference to individual parts numbers, controls, etc. The reader is advised to refer to the manufacturer's service instructions whenever possible, but first it is necessary to understand the purpose and function of these adjustments and this is the material covered in this chapter.

Vertical Sweep

The same vertical sweep circuit found in most monochrome TV sets is also used in the color TV receivers. The synch separator and oscillator should be familiar to the reader. Adjustment is critical only with respect to getting proper balancing of the three vertical controls; the vertical hold, height and linearity. Set the hold control first to obtain a steady picture. Then adjust the height and linearity controls for good linearity.

Variation of these two controls may cause the synchronization to be slightly off, which is indicated by pairing of lines and a slight vertical jitter. In a color TV set it is essential to get really good vertical linearity and, at the same time, jitter and pairing must be completely eliminated. Allow the picture to move from the bottom up and set the hold control so that it just snaps into place. Careful touch-up of the vertical hold control usually will eliminate jitter. Some color sets use vertical convergence circuits which also have some effect on the vertical linearity. Be sure to recheck vertical linearity after the convergence adjustments described at the end of this chapter are completed.

Because linearity is critical in many aspects of the color TV picture, most technicians use a crosshatch pattern consisting of vertical and horizontal lines as illustrated in Figure 15.1. Most color bar generators designed for the service technician also generate a crosshatch and a dot pattern. The vertical linearity and height controls should be adjusted with this pattern to make sure that the linearity is correct over the entire picture.

Color TV receivers use separate centering controls for the vertical and horizontal sweep section, as described in detail in Chapter 11. When the vertical centering control is varied over its two extreme positions, it should be possible to see both the top and the bottom of the picture and at that time the linearity can be checked again. The proper adjustment of vertical centering is to allow approximately ¼ inch of picture overlap at the top and bottom.

In Chapter 11 the pin cushioning correction circuits used in color TV receivers were discussed. Figure 11.4 shows an exaggerated example of pin cushioning while Figures 11.5 and 11.6 show typical dynamic cushioning correction circuits. The latter two illustrations show two adjustments which can be made to correct pin cushioning at the top and bottom of the picture. One of the adjustments is a potentiometer, controlling the amplitude of the pin cushion correction signal, while the other one is usually an inductance, determining the phase or the location in the picture, of the maximum effect of the pin cushioning correction signal. In most receivers the adjustments of the pin cushioning controls can be performed while watching the crosshatch raster on the screen. A few manufacturers, including Philco, provide a special test point to which the oscilloscope can be connected to view the pin cushion connection signal. The detailed manufacturers instructions should be followed for those circuits. In general, however, adjustment of the pin cushion correction controls will only require minor touch-ups and can therefore be performed by observing the raster.

Horizontal Sweep

Although the horizontal oscillator and AFC system used in most color receivers is the same as in standard black and white sets, the adjustment of the horizontal hold control is more critical. The reason is that in the color set the color killer, the burst gate, horizontal convergence, and the keyed AGC all depend on the horizontal flyback pulse for proper operation. If the horizontal sweep is even slightly out of phase with the incoming synch pulse, a defect due to faulty keying action in any of these sections can mask the true cause of trouble.

Whenever the horizontal hold control does not allow stable locking action at each of the channels received, both on color and monochrome, a thorough adjustment of the horizontal AFC is indicated. For those receivers using the phase detector type AFC, adjust the horizontal drive trimmer or discharge resistor at the grid of the output amplifier for maximum width without fold-over or bright vertical bars. Then carefully set the ringing coil tuning slug for good locking action with the hold control set at mid-point. Retouch the ringing coil setting until the picture moves horizontally but remains locked in while the hold control is turned to either extreme.

Difficulties in obtaining proper horizontal synchronization can be due to total loss or to reduction of amplitude of the horizontal synch pulses. This occurs occasionally when a special synch separator circuit, using noise cancellation, is used where an adjustment is provided for the amplitude of the noise cancellation signal. When such a control exists, it is necessary to connect the oscilloscope to the input of the horizontal oscillator phase detector to make sure that synchronizing pulses of the proper amplitude appear there at all times. Loss of synch signal is often due to overloading in the IF section and in the video detector section which, in turn, may be due to misadjustment of the AGC control and subsequent overloading. These conditions can be checked by connecting the oscilloscope to the input of the phase detector and observing the horizontal synch pulses.

The horizontal oscillator itself has usually only one or two, at the most three, adjustments which affect the frequency and thereby its stability. In many receivers the horizontal oscillator also contains a temperature sensitive or voltage sensitive resistor to stabilize the plate voltage of the oscillator. The operation of this component may have to be checked to eliminate it as the source of the instability. Troubleshooting the horizontal oscillator circuit in a color TV receiver is the same as in a monochrome set and will not be discussed here.

One of the indications of marginal horizontal locking action is the fact that the pictures weave, hum, or overload occurs for a few seconds after station switching or warm-up. Often the picture does not lock itself but some adjustment of contrast, hold, or the like, is required before it is synchronized. Careful adjustment of each of the various controls in the horizontal AFC circuit is the only cure for this defect.

Centering controls are also provided for the horizontal sweep. The adjustment of the horizontal centering potentiometer ordinarily presents no problem but must be performed after the horizontal synch circuit works properly on all channels. In some receivers it is possible to obtain horizontal synch over a broad range of horizontal hold control settings and this control also determines the apparent centering. Set the horizontal hold control to approximately midpoint and check the centering on a number of channels. The horizontal centering control should be adjusted so that each side of the raster overlaps the mask approximately ¼ inch and raster edges are not visible.

In Chapter 6 we have mentioned the necessity for regulation of the high voltage which is supplied to the ultor of the color picture tube. Figure 6.2 shows a typical automatic high voltage regulator circuit. Many recent receivers use a different H.V. control system in which the high voltage regulation is performed in the grid circuit of the horizontal output amplifier. Referring to Chapter 11 and Figures 11.2 and 11.3, we see the circuit connection of the H.V. regulation adjustment potentiometer.

Depending on the circuit used, individual manufacturers may recommend slightly differing procedures for adjustment of the high voltage. In all cases, the actual voltage at the picture tube ultor connection should be measured with the high voltage probe of the VTVM. Where a high voltage regulator tube is used, the manufacturers usually recommend adjustment of the high voltage potentiometer together with a cathode current measurement of the high voltage regulator tube. In the RCA models, for example, a special resistor is provided from the cathode of the regulator tube to B+. Connecting the VTVM across this resistor permits the measurement of the total tube current.

In receivers using the grid circuit of the horizontal output amplifier for voltage regulation, a jumper is often provided which permits inserting a milliameter into the cathode of that tube to measure the current while the H.V. regulating control is adjusted. The values of tube current and the actual ultor voltage will vary with different receivers, different color picture tubes and different circuits, but in each case the manufacturer's data should be carefully studied and the adjustments made accordingly. If only a touch-up is necessary, the high voltage should be adjusted with a black and white picture of average brightness and the voltage variation should be observed when a color picture is tuned or when the screen is completely black. In general the variation of ultor voltage expected between maximum and minimum currents should not exceed 10% of the total voltage.

The focus control is usually part of the high voltage section and is adjusted as part of the overall horizontal sweep adjustment procedure. The setting of the focus control is usually not very critical and its adjustment can be performed by observing the thickness of the horizontal scanning lines at both top and bottom and at the center of the screen. The left and right focus may require some compromise in that a small amount of defocusing may have to be tolerated at the sides in order to get the best focus over the largest area.

Color Synchronization

This portion of the adjustment procedure is not used in monochrome TV; additional test equipment is required for efficient alignment. The most important item of test equipment is a color bar generator of the type generating the multiple bar pattern. The VTVM found in most service organizations is also needed. Whenever available, a good oscilloscope with a frequency range up to 5 mHz will also come in very handy.

Basically, the color synchronizing circuits can be classified into two groups. One group contains an oscillator which is synchronized to the incoming color burst signal. The second type makes use of this burst itself and, by means of a crystal ringing circuit, produces the color reference signal directly from the synchronizing burst. Obviously the alignment and troubleshooting of the last type is simpler than the first. Both systems and actual receiver circuits of each were discussed in detail in Chapter 10. The alignment procedures for the color synchronizing circuits supplied by the manufacturers are sometimes designed to be done mechanically, without any clear explanation. Rather than simply follow a routine without understanding, the technician can do a better job by understanding the purpose of each step. The reader will be able to cope with defects and troubles which often spoil a prescribed routine and show the difference between a screwdriver mechanic and a skilled TV service man.

Alignment of the color synchronizing circuit is generally divided into two types of procedures. One type of alignment requires only the use of the VTVM and the availability of a color transmission or a color bar generator. This procedure lends itself to performance in the customers home and is therefore often referred to as field procedure. When this is not practical, a more detailed shop procedure, using a color bar generator, oscilloscope and VTVM will be required.

a) Field Procedure

Before beginning the field procedure it is necessary to check the setting of the color killer control. Tune in a station which transmits a color signal or connect the color bar generator to the antenna terminals and adjust the killer control so that full colors appear. Adjust the contrast and the chroma controls for medium level colors and set the tint control potentiometer to approximately mid-range. Depending on the circuitry used, ground either the input of the color burst amplifier or the input of the reactance control circuit, to let the 3.58 mHz oscillator run free. The VTVM should be used with a resistor isolating probe to minimize the capacity of its leads and should be connected to the grid of the 3.58 mHz oscillator. In this condition the burst transformer is tuned for minimum reading on the VTVM. The VTVM is then moved and the oscillator frequency is adjusted, usually by means of a tuning capacitor or a coil to produce a zero beat on the screen. This means that the color picture appears, at least momentarily, synchronized on the screen. If this can be achieved, the ground can be removed from the burst amplifier or from the reactance control circuit. As a final step, the VTVM is connected to the demodulator, at the point where the color synch signal is applied. The output transformer of the 3.58 mHz oscillator is tuned for maximum VTVM reading. After making the above adjustment, the tint control should be checked at both ends of the range and it should be possible to obtain both excessive greens and excessive purples in the picture. With the tint control set at the approximate center, the colors in the picture should appear natural. If a color bar generator has been used with a pattern like that of Figure 13.7, the third bar from the left should be red, the sixth and seventh bar should appear to have equal amplitudes of blue, and the tenth bar should be green. To provide a slight touch-up at this point, the 3.58 mHz oscillator output transformer can be adjusted slightly.

b) Shop Procedures

The shop procedure will vary somewhat between different types of receivers, different test points and different tubes and transistors. In general, however, a color bar generator, oscilloscope and VTVM are always used in the following sequence of steps:

The color bar generator is connected to the antenna terminals and the receiver is tuned for normal color reception. The color killer control is always turned to full color. The first step is to short-out the error voltage, or the burst amplifier input, or some other point which will prevent the oscillator from being controlled by the incoming color synch burst. The VTVM is connected to the 3.58 mHz oscillator, or to a suitable test point on the reactance control circuit, and the burst transformer is tuned for a minimum output at the VTVM. Next, the oscillator frequency control, a tuning capacitor or a coil, is adjusted, by observing the color pattern of the picture tube face until the color bars stand still or drift slowly. The remaining adjustments depend on the circuit but they are usually made with the oscilloscope, using a low capacitance probe, connected to the output of the matrix amplifiers or to the control grids of the color picture tube.

Figure 15.2 shows the correct oscilloscope waveforms obtained at the respective test points. It is necessary to move the oscilloscope probe back and forth between the three picture grids, or corresponding test points, while adjusting the output of the 3.58 mHz oscillator and the tunable phasing circuits, where they are used.

In receivers using the high level demodulator circuit described in Chapter 9 and illustrated in Figure 9.20, separate adjustments will be necessary for the R-Y and for the B-Y color synch signal. In this respect the adjustment of these two signals is nearly independent and can be made with reference to the oscilloscope presentation at the R-Y or the B-Y picture tube control grids respectively.

The alignment of a color synchronizing circuit using a crystal ringing circuit is relatively simple. It needs only a signal from a color TV broadcast station or from a color bar generator. Without shorting anything, each of the tuned circuits in the color synch chain is tuned for best color synch. When this is done, the only remaining adjustment is the balance of the demodulator and this is performed by connecting the VTVM to the center of the two diodes and adjusting the balance potentiometer to zero output on the VTVM. During the color synch

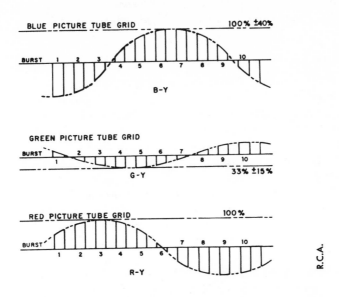

Figure 15.2—OSCILLOSCOPE WAVEFORMS AT THREE PICTURE TUBE GRIDS

section alignment it is good to keep the amplitude of the burst at a low input level and this is usually accomplished by either turning down the output amplitude of the color bar generator or, if a station is used, by detuning the receiver fine tuning control slightly after each adjustment. When the balance potentiometers are adjusted for zero VTVM output, the fine tuning control adjustment for best picture should not change the zero VTVM reading.

Color Killer Adjustments

Where a color killer potentiometer is used, its adjustment should be performed under two conditions. First tune in an unused channel in which some noise or snow appears on the screen. Set the color killer control so that color appears superimposed on the noise. Then change the setting of the killer control so that all color is removed from the screen and noise appears in black and white. Next, tune the weakest color TV signal and make sure that full colors are obtained. It may be necessary to switch back and forth between a weak color signal and between straight noise until the optimum setting is achieved. The operation of the color killer control can then also be verified by tuning to a black and white transmission and making sure that, even on strong signals, no color components flash through the screen.

Adjusting for Color Purity

All of the shadow mask type color picture tubes which use three electron guns also use some type of purity control device that adjusts the axial relationship of the three electron beams. As demonstrated in Chapters 5 and 6, the purity devices consist of PM assemblies which generate a magnetic field. The purity magnet is located on the neck of the picture tube, approximately halfway between the socket and the deflection yoke.

The detailed operation of the purity magnet has been described in Chapter 5, and the reader is referred to Figure 5.6 for the effects of the purity magnet tab adjustment. A detailed procedure which will work with practically all color TV receivers except those with picture tubes having the three electron guns in a single line, is presented below.

The purity adjustment should be made after the receiver has been on for a minimum of 10 to 15 minutes, when the set is operated with full brightness and with proper center convergence. The picture tube's screen should, if possible, be either facing to the north or to the south, to minimize the effects of the earth's magnetic field. The purity adjustment is made without picture information on the screen and therefore, where available, the service switch is turned to "raster only" or else the antenna terminals can be shorted-out, or an unused channel is tuned in. Purity adjustments are best made on red only and most manufacturers provide jumper terminals to disable the blue and green guns. The blue and green electron guns can also be disabled by turning down the respective screen voltage controls, but this means that these controls must be adjusted correctly after the purity set up is completed.

Loosen the yoke mounting clamps and move the deflection yoke as far away, backwards, from the screen as possible without displacing the convergence assembly. Observe the red screen and rotate the two tabs on the purity magnet assembly until the screen is uniformly red, at least in the center. Figure 15.3 shows poor red purity while Figure 15.4 shows an acceptable pattern. A low power microscope can be used to observe the position of the dots on the screen as indicated in Figures 5.6 and 15.5. Figure 15.5a shows the adjustments possible by the motion of the deflection yoke and their effect at the edges of the screen, while Figure 15.5b illustrates the adjustments for purity at the center of the screen. Rotating the entire purity magnet moves the dots in a circular path.

Spreading the tabs of the purity magnet increases the strength of the magnetic field and moves the dots in a radial direction, such as in the beam movement pattern of Figure 15.5b.

Adjust for center purity first and when that has been obtained correctly, move the yoke towards the screen of the picture tube while observing the entire screen. The yoke should be positioned for best overall red purity. Figure 15.6 shows the rear view of a typical color TV receiver on which the deflection yoke clamps, and the purity tabs behind them, can be clearly seen.

**Figure 15.5—EDGE AND CENTER
PURITY ADJUSTMENTS**

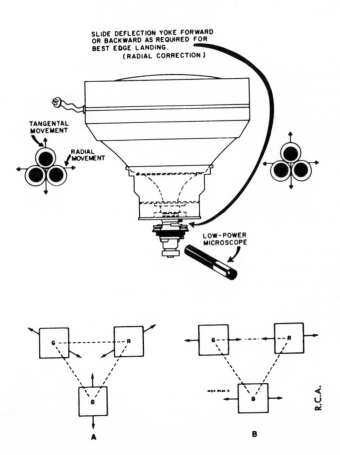

After the best possible red purity has been obtained, the blue and green electron guns can be activated again to observe the overall screen shading. The screen should be uniform white or light grey. It may be necessary to reconverge the electron beams at the center of the screen and repeat the purity adjustment.

Although the above instructions seem to indicate that purity adjustment is a tedious and difficult process, it is only a matter of a little practice until the technician will be able to perform this adjustment without any difficulty in a few minutes.

Convergence Alignment

It is feasible to adjust convergence with a black and white picture, especially if the picture is stationary as in the case of a test pattern. For the beginner and for an efficient adjustment method for the expert, a dot or crosshatch pattern is recommended since this permits exact observation of the convergence action all over the screen. The procedure outlined below makes use of a dot pattern generator or a similar unit capable of delivering a dot pattern in black and white. If a generator which delivers vertical and horizontal bars or a combination of both is available, this pattern can also be used. The relationship of the horizontal bars as a function of vertical convergence and linearity must be realized.

The vertical bars are indications of the horizontal sweep and convergence circuits.

1. Tune the receiver and generator to the same channel and connect the RF output of the generator to the antenna terminals.

2. Check the adjustment of vertical and horizontal hold, width, height and linearity controls. Turn the chroma control down and adjust the contrast and brightness for a good monochrome dot pattern. Adjust the fine tuning control and other controls until the dots appear clearly against a dark background.

3. For static convergence, set the dynamic controls to minimum and turn the PM poles for best center convergence. Use the blue positioning magnet assembly for horizontal movement of the blue beam.

4. With the center of the screen reasonably well converged look at a vertical column of dots, or a vertical line, if a crosshatch pattern is used, and adjust the controls for the red and the green horizontal convergence until the entire vertical line in the center is properly adjusted.

5. Observe a horizontal row of dots, first in the center of the screen, then at the top and the bottom, and adjust the controls which set the vertical convergence until these three areas are properly converged. It

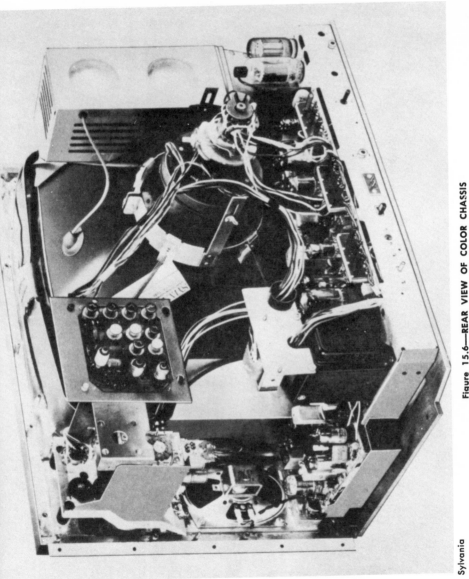

Figure 15.6—REAR VIEW OF COLOR CHASSIS

Sylvania

209

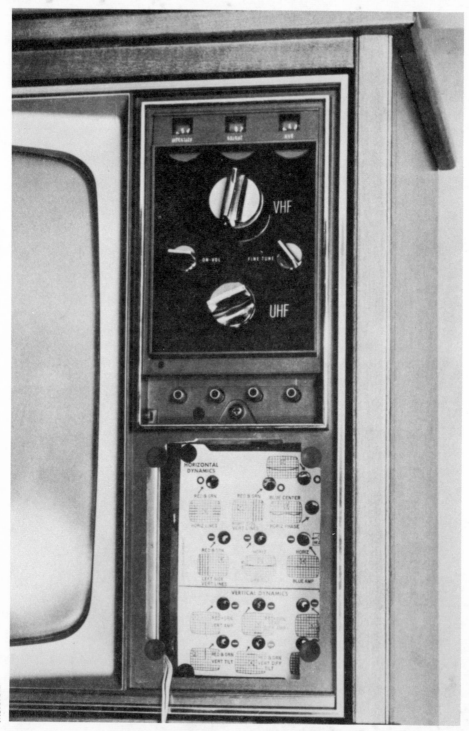

Figure 15.7—EXPOSED CONVERGENCE PANEL

may be necessary to readjust between the center line and the top and bottom line.

6. Observe vertical lines at the left and right edge of the screen and adjust the horizontal controls, back and forth until these two areas of the screen are properly aligned.

7. Minor touch-ups may be necessary and, occasionally, the blue magnet adjustment may have to be touched-up at this point to give good overall convergence.

Most manufacturers provide detailed instructions, often directly on the receiver. In Figure 15.7 a Motorola color TV receiver is shown with the speaker panel removed and the convergence adjustment exposed. Note that a crosshatch raster is shown there to aid the technician in adjustment. In many receivers of the RCA type a separate sub-assembly, shown in Figure 15.8 is accessible. It is usually mounted at the rear of the receiver and each control is clearly labeled. Detailed manufacturers

R.C.A.

Figure 15.8—CONVERGENCE SUBASSEMBLY

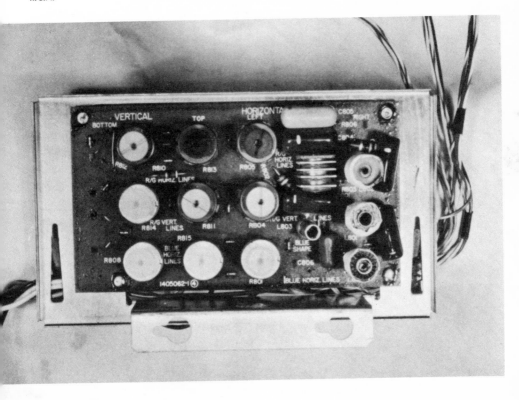

Figure 15.9—TYPICAL CONVERGENCE INSTRUCTIONS

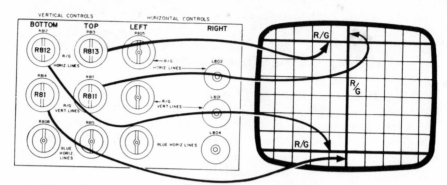

STEPS 3 AND 5 ADJUSTMENTS

Effect of R811, R812, R813 and R814 Adjustment

R.C.A.

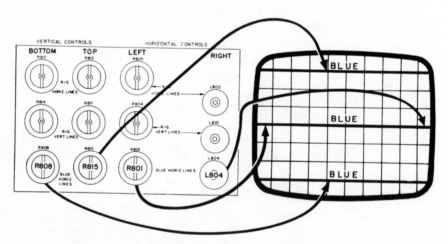

STEPS 6 AND 7 ADJUSTMENTS

Effect of R801, L804, R808 and R815 Adjustment

Figure 15.1—CROSSHATCH PATTERN, BADLY MISCONVERGED

Figure 15.3—POOR RED PURITY

Figure 15.4—GOOD RED PURITY

Figure 15.10–
–HORIZONTAL CONVERGENCE OFF

Figure 15.11–
–R-G VERTICAL CONVERGENCE OFF

Figure 15.12—GOOD CONVERGENCE

Figure 16.1—TINTED MONOCHROME

Figure 16.2—POOR PURITY, MONOCHROME

Figure 17.1—WEAK COLORS

Figure 17.4—LOSS OF COLOR SYNCH

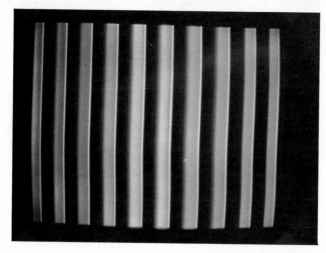

Figure 18.1—GOOD PICTURE

Figure 18.2—BLUE MISSING

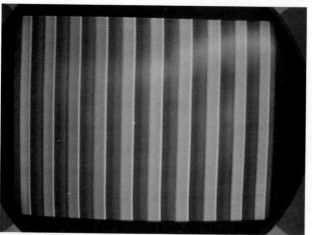

Figure 18.4—EXCESSIVE BRIGHTNESS

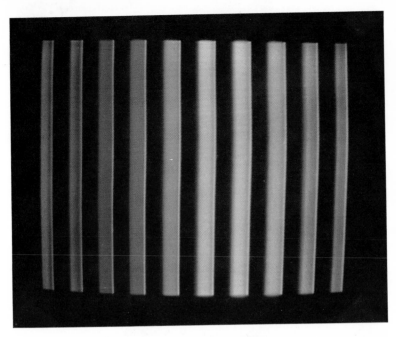

Figure 18.5—HUES WRONG

Figure 19.4—AUDIO INTERFERENCE

Figure 19.6—
-GHOST IN FULL COLOR

Figure 19.7—
GHOST IN MONOCHROME

Figure 19.8—
-GHOST IN SINGLE COLOR

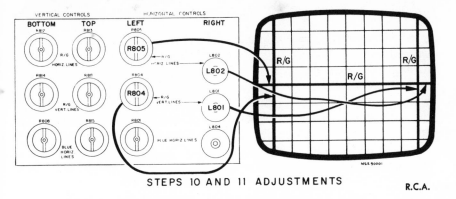

STEPS 10 AND 11 ADJUSTMENTS

R.C.A.

Effect of L801, R804, L802 and R805 Adjustment

CONVERGENCE

1. Use either crosshatch or dot pattern for center converge.

2. Converge center of screen with red, green, and blue magnets and the blue lateral magnet.

3. Adjust R811 and R814 for convergence (parallelism) of R/G vertical center line.

4. Readjust center convergence if necessary.

5. Adjust R812 to converge bottom R/G horizontal lines and R813 to converge top R/G horizontal lines at center line of screen.

6. Adjust R801 and L804 for straight horizontal blue center line.

7. Adjust R808 and R815 for uniform displacement of blue horizontal lines along center vertical lines.

8. Converge blue horizontal lines with R/G horizontal lines by adjusting the blue convergence magnet. Adjust red and green magnets if necessary.

9. Repeat steps 6 through 8 if necessary.

10. Adjust alternately L801 and R804 for right and left side convergence of R/G vertical lines.

11. Adjust alternately L802 and R805 for convergence of R/G horizontal center line.

12. Converge center of screen and repeat steps 10 and 11 if necessary.

13. Minor touch up adjustments may be made using the appropriate controls. If wide blue correction is necessary, loosen yoke and adjust wide blue correction screw. If wide blue correction is adjusted, purity must be rechecked.

213

instructions for this type of receiver are shown in Figure 15.9, together with the individual steps. It is reproduced here because it is very widely used and the technician will often have occasion to refer to it.

The typical effects of poor convergence can be seen in the color photograph of Figure 15.1 which shows a crosshatch pattern. Note that the blue lateral adjustment is obviously off because separate blue lines are seen throughout the horizontal scan. Where a yellow line appears this indicates that the red and green electron guns are properly adjusted. In this figure the horizontal adjustment for the R-G is correct but the vertical adjustment is off over the entire screen. The effects of poor convergence, primarily in the horizontal direction, is shown in the dot pattern of Figure 15.10. Note the white areas of overlap between the three colors, the overlap areas are between green and red which are yellow and the separate red, blue and green portions. In order to simplify convergence adjustments many technicians turn off the blue gun and perform the R-G convergence first. In Figure 15.11 the R-G vertical convergence problem is illustrated. The blue gun is off and the vertical convergence between the R-G controls must be adjusted until the dots are completely yellow.

To illustrate that good convergence can be achieved, we include Figure 15.12 which illustrates perfect convergence at least over an area of the screen. Note that the color of each of the dots is uniform throughout, there is no fringing and each dot is clearly defined.

All of the above procedures apply to shadow mask tubes which have the three electron guns and the color dots arranged in a triangle. For other color picture tubes the convergence procedure will be somewhat different and simpler. Although it may seem tedious and difficult, once the technician has performed the convergence adjustment a few times, the entire process will become so familiar that it will soon be as simple as setting the centering and linearity controls. From the above method it will become apparent why a dot or crosshatch generator is so useful for convergence adjustment. Only a really experienced technician should attempt to adjust convergence with a moving monochrome picture and only as a last resort when no generator is available.

16

TROUBLESHOOTING MONOCHROME OPERATION

The preceeding fifteen chapters have dealt with color TV principles, receiver circuitry, and the alignment of all adjustable components. Locating defects in color TV sets and repairing them is the subject of the next four chapters. Just as in the case of the alignment and circuit explanation chapters, we shall concentrate on troubleshooting those defects which are peculiar to color TV.

The technician who is familiar with black and white TV receivers will be fully aware of the standard tests which are used for practically all electronic equipment. A portion of the circuit is tested either with the power on and the equipment operating, or with the power off, tracing the signal paths or checking individual components. Component tests usually require only two types of test equipment. Tubes are checked best in the tube tester, while transistors, integrated circuits, resistors, coils, capacitors, etc. can almost all be tested by means of a good voltohmmeter or VTVM which contains an ohmmeter section. Detailed instructions for these tests are omitted here because we assume that the reader is thoroughly familiar with them. It will be necessary, however, to refer to such tasks as "Check the tubes in section," "check the transistors, measure resistance, check capacity, etc."

215

For troubleshooting color TV sets a 10 step procedure is recommended which will assure a professional job, the best way to customer satisfaction.

1. Verify the Symptoms

No matter what symptoms the customer may describe, the technician should always verify them because the untrained observer may often miss or misjudge different clues. The importance of this step will become apparent after the technician has listened to the descriptions customers give for such completely different troubles as sound in the picture, loss of horizontal synch or loss of color synch.

2. Localize the Defect to the Responsible Receiver Section

While it is possible to check all of the tubes, transistors, etc. in a color TV set, it is not practical and it certainly could waste a lot of time. The symptoms inevitably permit the localization of the defect to a certain functional receiver section which then allows trouble-shooting in a more efficient manner. If the picture appears satisfactory but no sound is obtained, for example, it would be a waste of time to troubleshoot any section other than the audio section. Similarly, the various color defects can be localized to individual circuits which speed up the troubleshooting procedure greatly.

3. Probable Trouble Source

In the following paragraphs and chapters we will try to point to the most likely source of the defect, based on the reliability of certain components as compared to others and based on the frequency with which certain circuits become defective. It is well known, for instance, that vacuum tubes are amongst the least reliable components in a color TV receiver and tube checking or tube substitution is therefore a fast way to find many defects.

4. Signal Tracing

If the trouble has not been cleared up by checking one of the likely sources it is usually necessary to trace the signal, such as the audio, video, color synch, etc., through a portion of the circuitry to find where it is lost or changed substantially. Signal tracing is a well known technique, dating back from the early days of radio and audio equipment, and is still of invaluable help in color TV troubles. Sometimes signal tracing leads directly to the defective component.

5. Component Testing

When the signal tracing method leads to a component or a group of components it is often necessary that each part be tested individually to find out which one is defective. Having eliminated the most likely trouble spots, component testing is usually confined to measuring the values of resistors, capacitors, trying substitutions, checking the continuity and performance of coils and transformers, etc. As mentioned above, the detailed procedure for performing these tests will not be repeated here, but the components likely to be defective will be pointed out.

6. Replacement or Repair

Most defective components must be replaced and only very few, such as poor wiring, can be repaired. The majority of components in a color TV receiver are standard, off-the-shelf, items but there are quite a few which are unique for a particular receiver model and for which an exact replacement part is required. In these cases, the exact replacement part must be obtained to assure proper operation.

7. Adjustment or Alignment

Many of the circuits in a color TV receiver will require adjustment or alignment after a new component has been installed. Chapters 13 through 15 include alignment instructions for each of the circuits in a color TV receiver which can be adjusted and aligned. In the following paragraphs reference will be made to these chapters and the reader is advised to go back to them for the particular alignment procedure.

8. Checking the Repair

After the adjustment and the alignment has been accomplished the technician should check the performance of the receiver to be absolutely sure that the defect has been really eliminated. Some defects occur only after the receiver has been warmed up for half an hour or longer and, if the repair and troubleshooting have permitted the chassis to cool down, it may then appear as if the defect has gone away. For this reason it is necessary to check and verify the repair thoroughly, allowing sufficient time for whatever thermal effect might be causing the trouble.

9. Checking Overall Performance

It often happens that a particular defect obscures the existence of other shortcomings of the receiver. Sometimes several symptoms occur simultaneously, one defect causing another, with only the ultimate one, such as no picture at all, becoming apparent. For this reason it should be standard practice for any professional technician to check the overall performance of the receiver carefully and make sure that good pictures and good sound are obtained on all channels used in the particular area. This process offers the technician an opportunity to install a better antenna, sell an antenna rotator, or perform some additional work in order to give the customer the full use of his television set. Checking the overall performance often saves call backs and avoids the customer complaint that, "While the picture is good now the sound is not."

10. Demonstrate the Receiver Performance to the Customer

No service call is ever completed until the customer has agreed that all troubles are cured and that the set really operates well. Before presenting the bill, the customer should be shown all channels, any remaining limitations should be demonstrated and explained, and the customer's satisfaction should be expressed. Many technicians find it worth while to affix a sticker with the date and the type of defect that was repaired somewhere on the chassis, as well as to leave their card with the customer for a possible return call.

One of the general problems which are applicable with both monochrome and color TV is the question of whether a particular defect can be repaired in the customer's home or whether the receiver has to be taken to the shop for servicing. This problem has been debated at great length for the past 15 or 20 years and no definite conclusion or firm guidelines have been evolved. In practice, each technician or service organization makes up their own rules, according to their ability and according to the availability of test equipment.

One basic rule, however, is followed by almost all professional technicians. This depends on the amount of time a particular defect will take. One has to make an estimate during step 1 or 2 of the above procedure as to whether this particular trouble can be repaired in less than an hour or if it will take longer. One has to decide whether this particular defect will require an overall, lengthy and careful, alignment of the receiver or whether a simple tube substitution or soldering in a component will be sufficient. As a general rule service calls which take longer than one hour should be converted into shop servicing. When a complete overall alignment or replacement of the color picture tube is required the shop servicing procedure is usually recommended. When removal to the shop has been decided on, the next question is whether the chassis only or the entire cabinet should be taken out. Here again, individual judgement can be guided by certain practical considerations such as the size of the cabinet, the weight of the cabinet, and the technician's ability to transport it undamaged, back and forth.

Technicians starting out often ask what is the minimum of test equipment that they can use for home calls and what is the best test equipment to have in the shop. It is difficult to supply a hard and fast answer to these questions but certain definite guidelines can be provided. For servicing color TV receivers in the home the following test equipment, in addition to hand tools, appears essential;

Color bar generator
VTVM or VOM
A box with tubes and other spare parts
Manufacturer's instructions for the particular set

The last item on this list is probably the most useful and the most essential. For this reason, experienced service organizations always require the customer to tell them the make and model of the receiver before a service call is scheduled.

As test equipment for the shop we would recommend, as a bare minimum, the following items:

Color bar generator
VTVM or VOM
Oscilloscope
Sweep generator
Tube tester
Transistor tester
Capacitor tester [the type that also tests inductance]

In addition, the well equipped shop will also contain a full supply of spare tubes, transistors, etc. All of the hand tools listed in Chapter 12 should normally be available in the shop.

Until the technician has gained sufficient experience and confidence in his ability to deal with any and every possible defect, we naturally also recommend that a copy of this book accompany him both on outside calls and for shop servicing.

It is always wise to work with the manufacturer's service data at hand or at least with a circuit diagram of the receiver. The trouble-shooting method presented here is not intended to replace the manufacturer's literature, but it will be very advantageous to use it together with any other data available.

For color TV receivers the possible color troubles can be broadly classified into four groups:

1. Defects apparent mostly in black and white operation.
2. Loss of color on color TV reception.
3. Wrong colors on color broadcasts.
4. Interference with color pictures and miscellaneous defects.

Chapters 16 through 19 will each cover one of these groups of troubles. Before starting to service a defective color receiver, determine which group of defects is indicated by the symptoms and look over that particular chapter. Refer to the color photographs associated with that chapter and then read up on the defects listed. A few minutes devoted to preparation will save hours of tedious tests and measurements and will avoid replacing, checking, and re-replacing perfectly good components.

While the vast majority of the defects are relatively simple component failures and will not present any real problem to the competent technician, there are a small percentage of defects which are very difficult to locate. Probably the most annoying of these is the intermittent defect. Intermittent defects are basically either of the thermal or of the mechanical, vibration or shock, type of origin. At the end of this book, in Chapter 19, some suggestions are provided which will help in locating both intermittent defects and a category of defects generally labled, "Impossible." The reader is advised not to refer to this portion of the book until he has absolutely assured himself that the defect does not fit into any of the other, readily identifiable, categories.

Tinted Monochrome Pictures

This defect is easy to recognize and fairly frequent, especially among new color sets where the owner was tempted to tamper with the color controls. On black and white broadcasts the entire screen area is colored in some manner, as shown in the color photo of Figure 16.1. This coloring need not be one of the primary screen colors since it is quite possible to get some mixture due to an unbalance of the color controls. However, the defects listed below and tests for them apply only to the case where the entire screen is tinted in the same color. When the coloring is apparent only in some spots, refer to the next heading.

The neutral shade white or gray which is desired for monochrome operation is dependent both on the three video signals and on the brightness conditions of the three electron guns. An unbalance in either of those two parameters can cause color tinting of the monochrome picture. Most receivers have controls which determine the color balance for the video and brightness circuits and, of course, these controls might be misadjusted. It is also possible that a defective component in either of these sections might cause the unbalance. The first recommended step is to determine in which circuit the defect lies.

1. Tune in a monochrome telecast or a color telecast with the chroma control set to minimum. Note the color which tints the picture. Turn the contrast control down until only the raster is visible. If necessary, switch to an unused channel.

2. Turn the brightness control up for a fairly bright raster. If this raster is tinted, reduce the brightness and adjust the background controls of the three electron guns for a neutral gray background. If the raster is not tinted and only a picture appears colored, the brightness circuits should not be disturbed since the trouble must be in the Y amplifier or in the matrix section.

3. If the background adjustment fails to bring the screen to a neutral shade, measure the DC voltages at the three screen and brightness controls. Measure the grid-to-cathode bias on each electron gun. It should be possible to set the three grid bias voltages to an equal value, adjust the three screen voltages to equal values and then touch up these controls to get a clean white raster. Breakdown in one of the bypass condensers, a defective bleeder resistor, or a similar failure may be responsible if correct DC voltages at the three kinescope guns cannot be obtained. A DC voltage check and finally an ohmmeter check will locate the defective part.

4. If the raster appears a neutral shade but the picture is still tinted, the defect is most likely in some portion of the color decoder or matrixing section. To make sure this is not due to misadjustment, touch-up the green and blue video gain controls, if available. A definite check on the video operation can be made by connecting the oscilloscope to each kinescope grid in turn and observing the blanking pulse amplitude at each grid. For a neutral gray picture the amplitude at all three grids should be equal.

5. Check the operation of the color killer circuit. Connect the oscilloscope to the output of each color demodulator in turn and make sure no video signal is present. Set the oscilloscope vertical gain to maximum for this test.

6. Since the Y signal goes to the kinescope cathode, there should be no video signal present at the three kinescope grids, but the signals at the three cathodes should be absolutely equal.

7. If it is found that one of the three matrix amplifiers draws more, or less, current, the amplifier tube may be weak, some component in

the amplifier circuit may be defective or the cathode bypass condensers may be open.

8. When the defect cannot be cured by the above steps, a complete resistance check of the matrixing network may be necessary. If one of the Y amplitude controls has opened up or changed substantially in resistance, a defect of this nature may be causing the tinting effect on the screen.

As a last resort, a complete signal tracing of the Y amplifier and the chroma section, decoder, and matrix may be required. Refer to Chapters 12 and 14 for the adjustment procedure of those receiver sections. The important thing to keep in mind is that the three cathode signals on the picture tube must be equal for neutral gray pictures, but the three electron gun beams should not have equal beam currents to produce a neutral gray, since the luminance values of each color are different and some correction must also be made for the different light efficiencies of the three colored phosphors.

Tinted Spots in the Monochrome Picture

The appearance of this defect is unmistakable and will be easily identified. Some portion of the screen, usually near the edges or corners, will appear tinted. The color may be varied over the area and may be anything at all within the range of the picture tube color gamut.

A typical defect of this type is shown in the color photograph of Figure 16.2. The first step is to reduce the contrast control until there is no picture visible and observe the contamination on a light gray raster.

This type of defect is invariably due to either incorrect alignment of the purity magnet or else due to stray magnetic fields. The reader is referred to Chapter 15 and the procedure given there for proper purity adjustment. In all cases where color contamination of the type shown in the Figure 16.2 appears, a complete purity magnet adjustment procedure should be tried first. The only exceptions to this rule are TV receivers which do not have an automatic, built-in degaussing system. In these cases stray magnetic fields may be causing the problem and degaussing would be indicated as the first measure. Most receivers, since 1965, have a built-in degaussing coil. Where it is not an automatic system it is only necessary to connect it, by means of a switch, to see the effects of the degaussing coil on the stray magnetic fields causing the color contamination. If the degaussing coil seems to have no effect, it is most likely that the degaussing coil itself has become defective. A simple DC check for resistance, followed by an AC voltage measurement when the degaussing current is supposed to flow through the coil, will verify this portion of the TV receiver operation.

If it is impossible to obtain good purity by the purity adjustment described in Chapter 15, and if the degaussing operation does not reduce the contamination, the problem can be due only to one of the following possible defects.

(a) An external magnetic field due to an extra, added loud-speaker, a large powerline transformer buried nearby, or some other machinery using magnets or transformers.

(b) A defective deflection yoke. The defect might be a very slight unbalance or a misalignment between the various windings. The only remedy or sure method of determining this defect would be substitution of a good deflection yoke.

(c) A defective convergence coil assembly. Usually such a defect is also manifest in poor convergence. If the three magnet coil assemblies shift their 120° spacing, color contamination as well as poor convergence could occur. This is an extremely rare condition and only substitution of a good part would prove this defect.

(d) The most expensive possibility is the picture tube itself. Very few instances are known in which the picture tube, after performing well for a while, develops color contamination. In some of the earlier 15-inch flat screen type tubes bombardment of the shadow mask could create dilated holes which then cause contamination. Again the only remedy is to try a new picture tube.

It should be kept in mind that few things are really pure and a minute amount of color contamination might have to be tolerated. The customer should be informed about such limitations and their negligible effect on the visible picture can be demonstrated.

Poor Focus

The poor focus condition occurs in monochrome receivers probably just as often as in color sets, but since the color picture tube operates somewhat differently and since troubleshooting focus defects in color is more complex, it is discussed here. Basically there are two different phenomena which can be called poor focus; the focusing can be poor or fuzzy over the entire screen or else the picture may be clear in the center and fuzzy only at the edges.

1. Poor Overall Focus

The first step in this case will be to check the adjustment of the focus control. For color picture tubes this control is located in the HV compartment and adjusts the DC potential on the focusing grids. The voltage for correct focusing will depend on the type of tube used. A simple way to check the operation of the focusing control is to connect a HV probe to the focusing element and meter the voltage as the potentiometer is varied. The focusing element is accessible at the tube socket and can be identified by the heavily insulated lead going to this socket pin.

If the voltage variation due to the HV potentiometer does not cover the range, check the ultor HV. Adjust the regulator control in the HV compartment for correct ultor voltage, then check the focus potential again. If only the focus voltage is low, replace the focus rectifier tube. Further checks to locate defective parts in the focus circuit can be made by ohmmeter measurements of the bleeder circuit in series with the focus control.

In the majority of cases poor overall focus is caused by some defect in the HV focus circuit. Occasionally, however, other troubles may be found. Excessive cathode current from any of the electron guns will cause "blooming" which will also have the appearance of poor focus.

As a last resort it is possible that the picture tube itself has become defective and cannot be focused correctly.

2. Poor Focus at the Edges

This phenomenon will be familiar from earlier monochrome picture tubes and poor deflection yokes which allowed a good focus adjustment either in the center or at the edges of the screen, but never in both places.

One of the favorite controls for the inexperienced customer to play with is the focus adjustment. Focus can be obtained, apparently, at several settings on some receivers and only the experienced technician, following manufacturer's instructions, can determine which focus setting is best. For these reasons, it is advisable in all cases of poor focus at the edges or center, to vary the focus control over its entire range. In some color TV receivers a compromise setting is normal, in which fairly good focus is obtained throughout most of the picture with the best focus at the center.

When a definite defect in the focusing has been verified the following steps are recommended:

1. Adjust brightness control for dim raster and vary the focus control for best focus.

2. Measure the ultor voltage and, if necessary, adjust the high voltage regulator control to obtain proper HV.

3. Measure the focus potential and adjust the control over the range. If the focusing voltage is correct, the defect must be elsewhere. Remember that correct focusing depends on correct ultor voltage and on the correct current from each electron gun.

4. Poor focusing sometimes accompanies other misadjustments or defects. Observe the picture carefully to make sure that good purity, correct convergence and proper color balance is obtained.

5. Occasionally some error during the setting up of the color picture tube and its components can cause poor focus. In some receivers the purity or convergence assembly can be mounted backward and perform correctly, but poor edge or center focus may result. The chances that the deflection yoke is defective are very slight if sufficient sweep and good linearity are observed. As a final possibility it should be mentioned that aging of the picture tube itself could give this effect. The owner should be advised that the tube is slowly going bad; a replacement will demonstrate this quickly.

Flat or Weak Pictures

This type of trouble is also found in monochrome sets and generally results either from incorrect IF alignment or a weak or defective video amplifier. It appears as in Figure 16.3. In color TV sets this defect is caused by the same trouble spots, but it is more noticeable, especially

Figure 16.3—WEAK OR FLAT MONOCHROME PICTURE

on color telecasts where it has the effect of distorting the gray scale. To locate the faulty section of the TV receiver, adjust the contrast control first to zero, then adjust the brightness control until the raster is a dim gray. Next turn the contrast control up until the black portions of the picture are really dark. Check to see if the white portions are now white or gray. As a more accurate check, connect the oscilloscope to the grid or cathode of each electron gun and observe the video signal. A good contrast video signal should appear as in Figure 16.4 with at least 60 volts amplitude peak to peak. Trace the Y signal through from the video detector to the kinescope and look for any sign of compression as the signal passes through an amplifier. In general the signal tracing procedure will readily locate this type of defect or at least localize it to a particular stage. Then conventional ohmmeter and voltage checks will ferret out the guilty component.

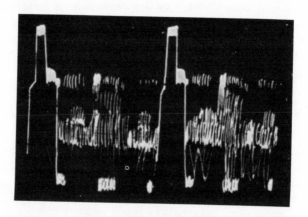

Figure 16.4—GOOD VIDEO SIGNAL, 60V PEAK TO PEAK

In tracing the video signal it should be kept in mind that on normal reception the output of the video detector should be at least 1 volt or more peak to peak. When the oscilloscope signal tracing does not indicate the trouble spot, an overall RF-IF alignment check should be made. Adjust the setting of the AGC control and check the AGC bias at the IF and RF stages. Excessively high bias can cause just this type of defect.

Another possible cause of flat monochrome pictures is the biasing at the kinescope or the DC restorer operation. Clipping of the peaks of the video signal can be due to high bias at the kinescope grids or else can be caused by incorrect bias on the DC restorer diode. Voltage and ohmmeter checks will quickly establish this type of defect.

Excessive Black Portions

While flat pictures are caused by the removal of the peaks of the video signal or by compression of the gray scale, the exact opposite can occur to give darkish pictures which appear to have too much contrast and little fine detail. Stretching of the video signal due to non-linearity in one of the video amplifiers is rare and can best be corrected by changing that tube. The other cause, incorrect IF alignment or overloading of the IF stages due to insufficient AGC bias is more frequent. A typical picture is shown in Figure 16.5 and while this looks bad in monochrome, it is still worse when a color telecast is being viewed. In color this defect is compounded by the phase shift and partial loss of some of the color information, resulting in an almost black, wrongly colored picture.

Figure 16.5—EXCESSIVE CONTRAST ON MONOCHROME

The first step is to adjust both contrast control and AGC control. If the contrast control has no effect on the signal at all, voltage measurements of the Y channel video amplifier should quickly locate the defect. If the AGC control cannot cut the signal down completely, measure the operating voltages, bias and the resistance in the AGC circuit. In this connection it should be mentioned that practically all color TV sets use keyed AGC, which requires a video signal and a flyback pulse on the AGC stage. Check these points if no AGC bias is available.

In color TV receivers using transistor or integrated circuit IF sections the AGC voltage polarity and operation is usually quite different from that used in tubes. Refer to Chapter 8, in particular Figure 8.11, for detailed information concerning the AGC operation and general IF circuits for transistor color TV receivers.

If adjustment of the contrast and the AGC control is capable of cutting the picture off, but not of changing its gruesome appearance, check the RF and IF alignment carefully. Measure the AGC bias at each stage which is normally controlled. Occasionally the RF chokes used in some leads open up, and while the AGC will then cut-off all other IF stages, one uncontrolled amplifier is sufficient to cause overloading.

Poor Convergence

In Chapters 5 and 6 the principles of convergence were discussed and in Chapter 15 the procedure for aligning convergence circuits was covered. From an understanding of these topics it becomes apparent that certain defects can develop which will give the appearance of poor convergence. When the customer complains of color fringing on monochrome reception, the best approach is to connect a dot or cross-hatch generator and perform the regular convergence procedure. During this procedure any defects in the circuit will be readily apparent.

If, for example, the green horizontal parabola control cannot bring the green electron beam in convergence at top and bottom, the defective part must definitely be in the green horizontal convergence section. Voltage measurements, observing the scope patterns and the effect of the controls, will quickly spot the defective component.

As mentioned in the previous chapter, absolutely perfect convergence over the entire screen is almost impossible to achieve, nor is it necessary. When viewed at the correct viewing distance, the individual horizontal lines as well as the color dots appear to merge and only the picture is visible. Similarly, the small color fringing effect due to slight imperfections in the convergence, become invisible at the normal viewing distance. This might be pointed out to the exacting customer who judges convergence by viewing TV with his face right on the screen.

17

TROUBLESHOOTING "NO COLOR" DEFECTS

If the TV receiver operates correctly on monochrome signals and even produces a good monochrome picture on color broadcasts, an adjustment of the chroma gain control should suffice to bring color into the picture. This chapter deals with the type of defects which can cause loss of the color component or else result in only a very weak, pale color picture. The case in which no color picture appears because the color synchronization section cannot be locked in is also discussed here.

No Color

Before starting an elaborate check or troubleshooting procedure in answer to the complaint that the set delivers no color pictures, be absolutely sure that color programs are really being received. In some areas the TV station might telecast network color programs in black and white because of the lack of color encoding equipment. A quick check of color performance can be made where the monochrome transmission includes a narrow color strip at the edge of the screen. Simply set the horizontal centering control until the strip is visible and observe its hue. General practice is to transmit a yellow-green strip, but local stations may use some other colors.

Only after ascertaining that no color can be displayed due to receiver trouble should a preliminary check be made. The following trouble-shooting procedure is suitable for all color receivers and will serve to localize the defect in a particular receiver section:

1. See if the filaments of all tubes in the receiver are operative.

2. Check the set briefly for loose plugs, cables, tubes partly out of the socket, and for other obvious mechanical defects.

3. Failing to find anything in the above steps connect the color bar generator to the antenna terminals and adjust both generator and receiver to the same unused TV channel.

4. Adjust the hue control and try varying the color synchronizing oscillator adjustment if accessible. If the latter adjustment gives some indication of colored bars running through the picture, the color synch section is suspect and the procedure given in the last paragraphs of this chapter should be followed.

5. Vary the chroma control to get some coloring. Next try the color killer and the green or blue video control or other gain controls in the color video or matrixing circuit.

6. Connect the oscilloscope to the three control grids of the picture tube and check for the presence of some video signal.

7. If the above step fails to show any signal in either channel, a wide band oscilloscope should be used to check the grids or input section of the two color demodulators for the presence of some signal. Both the color sub-carrier and the two color synch signals should be checked.

8. It may be necessary to trace the color sub-carrier and its side bands through to the video detector to locate the point where it is missing. Occasionally the color synchronizing burst is very weak and as a result the color killer circuit develops such a large bias at the chroma amplifier that this section is cut-off. Measure the chroma amplifier bias and disable the color killer temporarily.

9. Very weak color sub-carrier signal may be due either to IF misalignment, RF oscillator mistuning, or some misadjustment in the detector circuit. At the point where the color signal is separated from the Y signal, the 3.58 mHz sub-carrier should be at least several volts in amplitude.

10. If the color bar generator has an output at the color sub-

carrier frequency, it can be connected directly to the video detector; this will establish whether the signal is lost at the RF-IF or in the chroma channel.

As indicated above, the particular section in which the chroma signal disappears can be found. Once that section is located the defective part can be tracked down by simple resistance and voltage checks. Where pentode amplifiers are used, be sure to check the screen voltage as well as the screen and cathode bypass capacitor. An open capacitor can result in sufficient degeneration to lose the signal at that stage. The high level triode demodulators used in some receivers, connect directly to the kinescope grids without intervening video stages. To troubleshoot this circuit it is necessary to know all pertinent voltages and connections since both encoded and decoded signals are present in one stage. Be sure to check the 3.58 mHz trap alignment since shorted capacitors in such a trap can cause loss of color on the screen. If color signals are present up to the final stages, but no color appears at the screen, check each of the kinescope tube socket pins for the presence of color signals. Loose cable connections could cause such a defect as well.

If it appears that the loss of color signals might be due to failure in the color decoder circuit, check the output of each decoder section in turn with the oscilloscope. No video signal output at both demodulators when there is a 3.58 mHz signal with side bands present at the input to the demodulators means that there is no color reference signal being supplied. To verify this, measure the gating or suppressor grid bias on each demodulator if pentodes are used. In a properly operating circuit the DC bias on the suppressor grids should not be sufficient to cut the tube off completely. Where solid state demodulators are used the presence of the color reference signal can be verified by either the wideband oscilloscope or by using a diode detector probe and checking with the VTVM.

Weak Color

A typical pale picture with insufficient chroma content is shown in Figure 17.1. Note that all colors appear more like pastel shades than fully saturated vertical bars, and the general appearance is that of a lightly colored black and white photograph. On properly operating receivers the adjustment of the chroma gain control will correct this appearance promptly.

Before going into a detailed troubleshooting procedure, the operation of the receiver on black and white color telecasts should be checked first. Only if black and white pictures are received with satisfactory contrast, should the color section be suspected. If the black and white pictures also appear weak, the reader is referred to Chapter 16 and the troubleshooting hints for flat or weak pictures. Assuming that monochrome reception is good, the defect is bound to be in one or the other of the color sections. Because all three colors will appear equally weak, the most likely source of the defect is the chroma or color IF section which provides all three color signals. Checking the tubes in this stage, as well as the various voltages, particularly the color killer bias and its adjustment, will be the first step in the troubleshooting procedure. If these preliminary steps do not locate the trouble the following procedure is recommended:

1.   Vary chroma control and observe its effect. If the paleness does not change into a pure monochrome picture at zero chroma gain, the defect is most likely in that control or in the circuit associated with it. A typical example of this is shown in Figure 17.2. Here two controls affect the gain of both the Y and chroma channel. One potentiometer is generally labelled the contrast control and the brightness control is a separate potentiometer in the bandpass or chroma amplifier itself. A defect in one of the ganged potentiometers could give the impression of contrast control action with the chroma gain very low. If the chroma itself is defective a simple resistance check will show it at once.

Figure 17.2—CONTRAST AND BRIGHTNESS CONTROLS

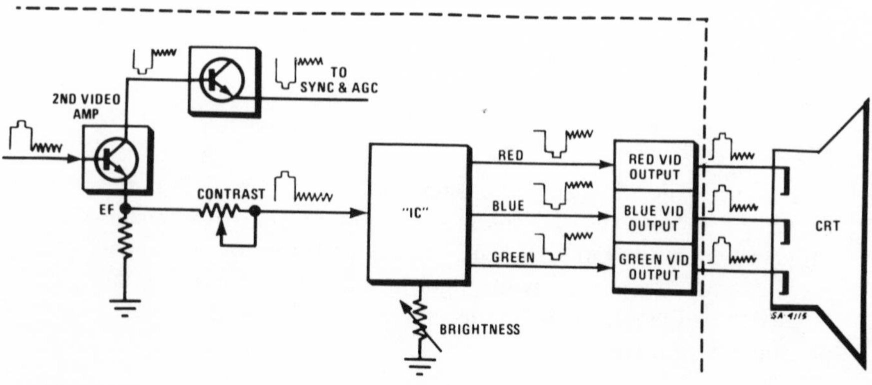

Motorola

2. Assuming the chroma gain control can cut the color off, but cannot deliver sufficient chromaticity for a normal picture, the next step will be to check the video signal output from the two demodulators. Sufficient signal at those points indicates that the defect must be in a stage between the demodulator and the kinescope. It is unlikely that weak colors would result if one of the three final color channels or one of the two color difference matrixing amplifiers were weak. In such a case the colors would be wrong, but not weak.

3. If the chroma gain control has no effect whatever on the chromaticity of the picture, the circuit of the gain control, the chroma amplifier stage itself and its operating voltages should be carefully checked.

4. Other factors which can cause weak colors regardless of chroma control setting are: excessive bias due to the color killer circuit, or excessive bias due to a chroma AGC system or some similar defect. High color killer bias has been found on some network color programs when the amplitude of the color synchronizing burst is somewhat low, causing the color killer which operates on the color synch burst only, to develop considerable bias. This trouble should be corrected at the transmitter but since the technician usually has little control over it, the next best thing is to reduce the color killer bias. This can be done either by changing the signal level at the color burst input or else by adding a voltage divider which reduces the developed bias slightly.

5. The RF and IF sections are not likely trouble sources for weak chroma defects as long as the overall monochrome picture appears correct. Only if the monochrome picture looks as if there is no fine detail and the original range of the contrast control appears reduced, only then should the IF section be suspected. It is possible that due to a weak amplifier or a defective component, the overall response curve of the RF and IF sections has been altered so as to give very little gain at the lower IF frequencies. A typical defective overall response curve of this type is shown in Figure 17.3. Note that the color as well as the monochrome video signal carrying the fine detail are all below the 50 per cent amplitude of the video carrier frequency.

Figure 17.3—WRONG OVERALL RESPONSE—LOSS OF SUB-CARRIER

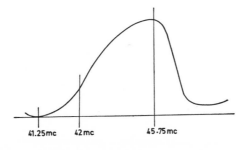

41.25mc   42mc   45.75mc

Color Synchronizing Trouble

From the first few chapters of this book it will be apparent that no color pictures are possible at all when the color synchronizing signal is lost. If the color reference signals supplied to the two demodulators are slightly out of phase with the transmitted color synchronizing burst, then the colors will be wrong. If the color reference signals are completely off in frequency as compared with the transmitted color burst, then there can be no color signal demodulated and the picture cannot contain any colors at all. When a color synchronizing defect is considered, either there is no color at all or else a series of red, green and blue bars travel over the picture like the one shown in Figure 17.4.

In this instance the color reference signals are approaching the frequency of the color burst. As the bars get wider, the frequency gets closer to the correct one until finally the entire screen is filled with color. While it is desirable to perform most of this with a station signal, the absence of a color telecast does not exclude troubleshooting color synchronizing defects. It is possible to use a 3.58 mHz signal generator as substitute for the burst, especially in crystal ringing systems. A color bar generator which supplies a standard color burst can also be used in place of a station signal.

To determine quickly whether the color oscillator is slightly out of color synchronism or whether the color AFC system is completely defective, adjust the accessible color phase or tint control first. Next try varying the color frequency controls or tuning networks located on the chassis. If none of these adjustments have any effect at all, try shorting the color killer bias, adjust for more contrast, chroma and less AGC bias in order to get the strongest possible color burst. If it is possible to synchronize and get a color picture with excessive contrast or chroma only, then the color oscillator is probably good, but the color synchronizing burst is insufficient. Replace the burst gating amplifier, the chroma amplifier, and the phase detector. If tube replacement does not cure this trouble, the chassis must be removed from the cabinet and a full troubleshooting procedure as shown below is required. In cases where there is no change in color synch, regardless of any of the above steps, the most likely defect is in the color oscillator itself. Replace the oscillator and control tubes as well as the phasing amplifier, and, in turn, each of the color demodulator tubes. In receivers using a crystal ringing circuit one of the amplifier-limiter stages may be defective. Try changing these tubes first. In solid state receivers the chassis will have to be removed for troubleshooting as shown below. The procedure outlined here includes any type of defect which can cause absence of color picture due to loss of color synchronization. The completion of the repair should be followed by a realignment of all adjustments as outlined in Chapter 15.

1. Using a wideband oscilloscope and the Color Bar Generator, connect the low capacity probe to the gating or control input of each demodulator stage, or to some other point where the color reference signal is applied. If a substantial 3.58 mHz signal is observed, the color oscillator is apparently operating and the defect is in the synchronizing circuit. In cases where a crystal ringing system is used, presence of a signal of approximately the right frequency could be due to regeneration or self oscillation. It does not automatically mean that the crystal ringing circuit and limiter stages are all good. To make sure that no regeneration exists, short each limiter input to ground and finally short the burst gate input to ground. This should remove the signal completely, otherwise there is some oscillation present.

2. When no color reference signal is observed at the demodulator, trace the path of the signal back to the color oscillator or the first limiter amplifier in a crystal ringing system. Using the scope detector probe, it will be possible to locate the point at which the signal is lost. Voltage and resistance checks at that stage will then pinpoint the defective part. In oscillators using a control tube, be sure to check all error voltages since excessive bias may cut the oscillator off. Crystals, either in the oscillator or ringing system may become defective. When checking with an ohmmeter use caution not to burn out the crystal with excessive DC voltage.

3. Assuming the color oscillator works, but synchronization is not possible, or else there is no sinewave present in a crystal ringing system, connect the wideband scope to the input grid of the color burst gate. The VTVM is less useful here, but since some manufacturers give voltage readings for the input of the burst gate with correct signals, such a reading can be taken as indication. If no burst signal is present at the burst gating input, the defect must be in the preceding stages. Check the output of the chroma amplifier from which the burst is obtained. If no chroma signal is present, follow the troubleshooting procedure given in the first paragraph of this chapter.

4. If the color synch burst is apparent at the input to the burst gate and not present at the output, the defect must be in the gating amplifier. First check the operating DC voltages and then observe the AC operation as follows:

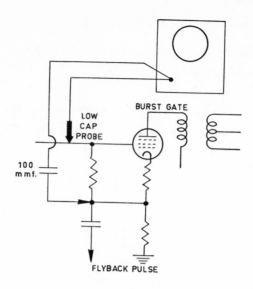

Figure 17.5—SCOPE CONNECTIONS FOR VIEWING COLOR BURST AND GATING PULSE

(a)   Connect wide band scope lead to the input of the burst amplifier through a low capacity probe and observe the color burst and horizontal synchronizing pulse.

(b)   Leave scope connected as in (a) but add a 100 mmf capacitor from the point at which the gating pulse is applied to the burst amplifier as shown in Figure 17.5.

(c)   Now the scope should show that the gating pulse coincides with the color burst as in Figure 17.6. Note that here a negative pulse overcomes the cathode bias. Otherwise the color burst cannot pass the gating amplifier.

Figure 17.6—SCOPE PICTURE OF BOTH BURST AND GATING PULSE

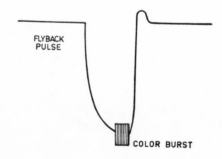

It is possible that due to some defect the horizontal pulse actuating the gating circuit is reduced in amplitude so that it cannot overcome the bias on the gating circuit to let the color burst pass. Another trouble is a delay in the gating pulse due to some breakdown, causing the burst to be partly lost. Be sure to check the amplitude of the gating pulse as well as the DC voltage on the burst amplifier if the color burst appears lost at that point.

5. If the color synch burst at the output of the gating amplifier is of the correct amplitude and the color oscillator is operating, the only reason for lack of synchronization can be in the phase detector circuit. Before following a detailed check of signals at each point, a quick resistance check of this circuit is suggested. Try aligning both primary and secondary of the phase detector transformer as outlined in Chapter 15. Finally make a point by point scope test for correct signals at the input and output of the phase detector circuit. Note that the local oscillator feedback signal should have more amplitude than the burst. This latter signal should appear with equal amplitude at the two sides of the transformer secondary. Remember also that the error voltage needs a long time constant filter (Figure 10.6), otherwise there will be "hunting" in the AFC action.

6. In a ringing crystal circuit the crystal may have to be replaced. Not every crystal resonant at 3.58 mHz is satisfactory for this job. Only an exact replacement, using the same cut, as specified by the manufacturer, should be installed. In general, defects in this system are easier to locate since the color burst signal need only be traced through.

Most color synchronizing defects will not require the full troubleshooting procedure listed above but will be fairly simple and routine. Realignment of the entire color synchronizing sections as described in Chapter 15 is recommended after any repairs have been completed.

18

WRONG
COLOR DEFECTS

One of the great problems in servicing TV receivers is to determine the correct coloring of a picture. As we know from the earlier chapters on colorimetry, any color is given by three criteria—the brightness, hue and saturation coordinates or the amount of red, green, and blue light. In either event, colors appear slightly different to each individual observer and, in addition, the surrounding colors will have some influence on the actual appearance of a particular shade. When we add to these factors the varied room lighting, color scheme of the home, range of brightness setting and the difference in lighting between studio or outdoor scenes, the difficulty of judging a color TV picture will become apparent. When a customer complains that his color pictures appear unnatural this may be due to any one of the external factors mentioned above or it may even be due to a simple slight misadjustment in one of the operating controls on the receiver. As still another possibility, wrong coloring may be due to a receiver defect. In this chapter coloring effects will be illustrated which are frequently due to receiver defects, but a simple complaint of "wrong colors" may mean anything from turning down the contrast control to a complete realignment of the color synchronizing section.

Before deciding that a particular defect is really a color trouble discussed in this chapter, set the receiver for a good monochrome picture. There should be ample brightness, good contrast and a neutral gray shading of the raster. Be sure that convergence and purity are correct by observing fine detail for convergence and the overall background for any coloring. Only if the monochrome operation of the receiver is satisfactory and color pictures can also be received correctly in black and white, should a defect be classified as wrong coloring.

To allow the serviceman some kind of indication, the color signal should contain known and easily identifiable colors. Moving scenes, especially studio pictures containing such artificial lighting as colored spotlights are not suitable for comparison tests. The best station signal for correcting color defects is a color TV test pattern which is occasionally used during morning hours. Many stations transmit a color bar signal as a test pattern. The color bar generator which is part of the serviceman's color test equipment provides the most suitable signal for color tests.

Because the color test pattern due to the bar generator is exactly specified, and each of the colors are known, it is possible to diagnose wrong color defects quickly with this pattern. Figure 18.1 shows the correct colors for the standard color bar generator test pattern. Note that the extreme left bar is yellow, followed by orange, followed by pure red. At the extreme right the vertical bar is pure green, at the left is a blueish green, followed by cyan, followed by two bars of equal intensity blue. The exact color range for the ten bars is described in Figure 14.7 and most color bar generators also contain this information either on the control panel or on the cover. Defects in the color TV receiver which result in wrong colors usually affect one or more of the primary colors. The color bar pattern contains the primary colors as well as some of their mixtures. The defect usually becomes clearly identifiable because the bar, corresponding to the primary color that is defective, will be weak, missing or excessively strong. Typical of this is the color bar pattern of Figure 18.2 in which a blue bar is missing. Note that in this pattern the yellow, orange and red, as well as the green, appear correct but all those colors which contain a portion of blue are wrong. The two bars which are supposed to be clear blue, are missing altogether.

Wrong colors are usually due to a defect in one or more of the sections carrying the chroma signal. This includes the chroma amplifier, both demodulators, matrixing circuits, color video amplifiers and the color synchronizing section. Misadjustment of any of the controls in these sections as well as in the brightness circuits of the color picture tube can also give wrong color effects. To get a closer indication of where a particular type of trouble usually originates, the possible wrong color defects are divided into the following groups according to their appearance. Each group is then discussed separately. This allows the reader to decide which type of color defect is apparent and then follow the troubleshooting procedure outlined for that defect.

1.   Lack of red. This is apparent in white areas being colored greenish, yellow areas appearing green, and purple areas seeming bluish.

2.   Lack of green. Here the purple tones predominate. Yellow shades become reddish.

3.   Lack of blue. For this color defect the predominant color will be yellow, a mixture of red and green. Deep blue areas will be black and the sky greenish. Figure 18.2 illustrates this defect.

4.   Dominance of red. This differs from the above three defects in that the dominant color is one of the primary colors, red in this instance.

5.   Dominance of green. As in the instance of red, green tinting of the entire picture is apparent.

6.   Dominance of blue. This is often not as clearly apparent since a small increase in blue is not too noticeable, especially for viewers used to the bluish cast of a monochrome picture tube.

7.   All hues wrong. This can result in purple flesh tones, green skies, and the like. In general such a picture is an indication of wrong color phase. A little practice will permit the service man to recognize this defect at once.

8.   Too much color. The colors appear overly rich and none of the pale, delicate shades are visible. In the color bar generator pattern this usually appears as "Blooming" of the color bars containing mixtures.

9.   Purple-greenish coloring. In this picture the predominant colors are purple and greenish; the orange or yellow shades are missing.

Before discussing each of these color defects individually a few general tests should be considered by the serviceman. Referring to Figure 18.3, the chroma producing sections are outlined and, if the monochrome picture is correct, the defect will probably be located in the sections shown. In some receivers the functions of two or three of these sections may be combined in a single tube, but this does not change the troubleshooting procedure. One of the general tests sug-

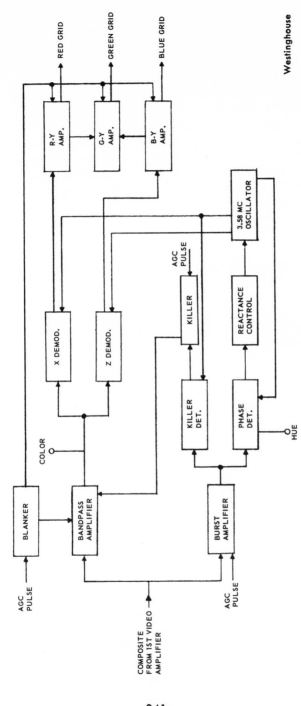

Figure 18.3—BLOCK DIAGRAM OF CHROMA SECTION

Westinghouse

241

gested before following the specific color defect tracing methods below is to adjust all receiver controls carefully. The IF and RF alignment need not be touched up if a good monochrome picture is obtained and only the brightness, background, contrast, and color controls should be adjusted. In Chapters 14 and 15 detailed procedures are presented for these adjustments; most of them can be performed with the chassis in the cabinet.

If the overall adjustment procedure does not correct the color defect, the chassis may have to be removed and another general test can be performed. Measure B-plus voltages in the color demodulator section and measure the bias voltages at the three kinescope grids.

When comparing the picture on the screen with the known color bar test pattern it is important that the hue or tint control be set properly. If it is impossible to adjust the color phase so that at least one of the ten color bars is at the correct color, the defect is most likely in the color synchronizing section. Only if at least one or two of the color bars on the screen correspond to the correct colors, is it possible to find the defect for the particular deficient color or color combinations. By analyzing the color bar pattern, the technician will be able to determine very quickly which particular colors are wrong. Remember that the yellow, orange, magenta, purple and cyan are each mixtures composed of two primary colors. To check whether the wrong color may be due to misadjustment of the background or screen grid controls of the color picture tube, advance the brightness control until excessive brightness, as in the illustration of Figure 18.4, is obtained. Now the spaces between the color bars will be a grayish shade. Examine this gray to make sure that it is a neutral color and does not contain a particular primary color. If the background controls are improperly set, this will be apparent on monochrome reception, but adjusting for excessive brightness on the color bar pattern is a quick way of determining it.

1. Lack of Red

Loss of the red and yellow bars indicates that the correct color balance is missing due to insufficient red video signal. Actually, misadjustment of the background controls could produce a similar appearance, except for the fact that the pure gray areas would be colored while loss of the red video signal does not affect the background at all. To be sure, however, the overall check of adjustments suggested above was made to get a neutral gray background.

In receivers having separate color video amplifiers the red video channel is the first suspected section. First try the red video gain control and replace tubes in this section. If this does not remedy the defect, the chassis will have to be removed from the cabinet. With the oscilloscope a simple gain measurement can be made and the red video output can be checked directly at the kinescope grid. In receivers using the R-Y, B-Y decoding system, the defect is often in the demodulator or else in a phase inversion circuit, losing or attenuating the red difference signal somewhat. Where a high level demodulator circuit is used, the defect may either be in the demodulator circuit or else in the video output network going to the red kinescope grid. Bandpass networks or 3.58 mHz traps have a tendency to short or open, resulting in loss of one of the color video signals. In general it is safe to assume that this defect can be localized in the red video, matrix, or R-Y demodulator section. Simple scope circuit tracing will locate the point at which the red video signal is missing.

2. Lack of Green

The same troubleshooting procedure as for the lack of red above is indicated here. One word of caution for the green color difference signal is that in most receivers using this decoding system, the red and blue difference signals are demodulated first. The green signal is obtained by mixing the R-Y and B-Y together. For this reason the green matrixing network should be given extra attention in troubleshooting. Defects in the R-Y or B-Y signals fed to the green matrix could cause a weakened green signal at the kinescope.

3. Lack of Blue

This defect occasionally is quite slight and may not be noticeable until it gets quite bad. The reason for this is the relatively low visibility of the blue light. Thus a blue video amplifier may deteriorate slowly without the customers realizing it until some day a bright blue area appears as black. The color bar pattern of Figure 18.2 shows this defect clearly. All of the troubleshooting procedures given above for lack of red in paragraph 1 apply in this instance but here the B-Y channels will be the major suspects.

4. Dominance of Red, Green or Blue

The defects described in the preceding three paragraphs have been due to an insufficient amount of red, green or blue color difference signals. It can also happen that one of the three primary colors is particularly strong and dominates the picture. This type of defect is generally due either to wrong color difference signal amplitudes or wrong adjustment of the three color picture tube screen grid controls. Before starting the troubleshooting procedure, observe a monochrome picture, and then the dim gray raster without the picture, to make sure that the color balance is correct. Check the background controls and the adjustment of the green and blue video drive controls.

With the oscilloscope connected at the three picture tube grids and the color bar generator connected to the antenna terminals the waveforms and various amplitudes shown in Figure 15.2 should be obtained. The color which dominates the picture will have a considerably larger amplitude at its picture tube grid. It will then be possible to trace back from the color picture grid to the point at which the amplitude becomes excessive. Excessive amplitudes mean that there is most likely excessive gain in some stage. This type of defect is usually due to a defective resistor in a matrixing circuit, a shorted capacitor in a cathode bias circuit or a similar component or wiring defect which could cause excessive amplification. Occasionally the defect is in the output circuit of the Y amplifier, and one of the three picture tube cathodes receives too much Y signal. When the stage is located at which the excessive gain is produced, voltage and resistor measurements will invariably locate the defective component.

In a normal color picture the effect of the wrong color phase will be that all of the hues in the picture are off. Faces may seem greenish or purplish, a blue sky may appear greenish, etc. In the color bar pattern of Figure 18.5 the effect of a misadjustment of the color phase controls is apparent when the individual color bars are compared with those of Figure 18.1. Note that, in effect, the hues of the bars have been shifted to the right. Thus the last color bar at the right is no longer green but a blueish green while the first color bar at the left is now green and the second one is yellow. This indicates phase shift in one direction. If the hue control were badly misadjusted in the other direction the effect would be the reverse and the last color bar at the right would be yellow and the first one at the left would be orange. Depending on the extent of the misadjustment, the color bar pattern may be displaced even more. In any event, this type of defect is invariably due to incorrect color phase at the demodulators. To produce incorrect color phase at the demodulators the color synch section is usually the first suspect. This is particularly true if all hues are wrong, as in the color bar pattern of Figure 18.5. Both the R-Y and B-Y or X and Z color synch signals are out of phase with the color reference burst. The following troubleshooting procedure is recommended:

(a)   Vary color phase or hue control. If this has no effect, or cannot completely correct the defect, change the phase amplifier in the color synch section. The chassis should be removed from the cabinet if this does not help.

(b)   Examine the color synchronizing section, phase detector, oscillator control circuits and the input to the demodulator stages for obvious mechanical defects such as melted wax, charred resistors, and so forth.

(c)   Perform the complete color synch alignment as shown in Chapter 15.

(d)   If some adjustments cannot be made or correct phasing cannot be obtained, the defective part will be located when some coils do not tune or shorted capacitors and the like are found.

5. Dim Red and Blue Tones, No Brightness

Although this defect is classified as wrong colors it is also apparent on monochrome reception. The Y or brightness signal is missing or is very weak and therefore the color components predominate. To verify that this is the case, tune in a monochrome reception or, when not available, set the color killer control to cut-off the chroma bandpass amplifier and it will be seen that no picture at all is now visible. The defect will invariably be located in the Y amplifier or the delay line. In some instances a very weak picture is possible, but the Y signal is definitely not of the correct amplitude.

The first step in this type of defect requires replacement of the brightness video amplifier tubes. If this does not clear up the trouble, the chassis will have to be removed from the cabinet and a more detailed procedure of locating the defect is suggested.

Troubleshooting involves nothing more than signal tracing the brightness signal with the oscilloscope from the video detector to the kinescope or to the matrixing network. By this method the point of signal loss will quickly indicate the defective component by means of a voltage and resistance check.

6. Too Much Saturation

When the color pictures appear excessively rich, and pastel cr subtle colors cannot be achieved this is classified as too much saturation. The chroma gain control or color gain control should be turned down to reduce the saturation and to obtain regular colors in the picture. When this is not possible, try reducing the picture amplitude by disconnecting one of the antenna leads or by partially shorting the antenna with a 50 ohm resistor. Excessive saturation can be due to excessive gain either in those stages carrying both the brightness and color information or in the color channel itself.

For this reason the first step after this defect has been verified is to adjust the contrast and brightness controls as well as the fine tuning and AGC controls for a good monochrome picture. Next, try a color picture. If it appears impossible to reduce the chroma gain control to get a monochrome picture, the chassis will have to be removed from the cabinet and a detailed voltage and resistance check of the chroma gain potentiometer and its associated circuit are required. Occasionally the chroma gain control appears to function correctly, but a good chroma setting is not feasible because after a few seconds the picture again looks oversaturated or else gets too pale. A gassy, intermittent chroma amplifier tube will cause this type of defect.

Another possible cause of excessive chroma gain could be the chroma AGC system in those receivers using more than one stage of chroma amplification. Again this defect can be located by simple voltage and resistance checks.

7. Color Phase Drift

This defect is most frequently observed in conjunction with warm-up or some other temperature variation. It consists of a change in the hues of a color picture in such a manner that it gradually passes through the correct hue combination, then goes through all possible wrong coloring and finally passes again through the right hues. This change in colors may occur so slowly that it is hardly noticeable or else it may occur rapidly. In some receivers this defect may only be present during the first few minutes of operation, after which an adjustment of the hue control is sufficient to lock it in to the right color phase. This defect may be very difficult to locate and usually requires considerable troubleshooting and testing in the shop. As the first step, each of the tubes in the color synch section can be replaced, but unless heater-to-cathode leakage is found, changing tubes will not usually cure this trouble. From the description of the defect we can immediately deduce that it originates in the color synch and phase control section. Depending on the speed with which it changes hues it may be possible to observe the drift in 3.58 mHz phasing with the oscilloscope.

First consider the case where color phase drifts only during warm-up time. A slow-heating oscillator control tube may be the trouble or else a defective electrolytic capacitor may be self-healing with the heat rise. If one of the capacitors in the error voltage filter network is leaky, the error voltage build-up may be very slow. An open resistor or high resistance lead in this circuit could also cause this defect. Even poor solder joints which require heat to make good contact can cause this type of defect.

Where the changing of hues in the picture is fairly rapid, one of the error voltage filter capacitors may be open, causing the APC system to "hunt." Open electrolytic decoupling capacitors may cause 60-cycle modulation of the reactance circuit, and this may appear as hue shifting at some beat rate between the transmitted vertical frequency and the AC line.

Hunting of the color APC system can be due to some other defect in the phase detector circuit, such as unbalance, reduced emission from one of the diodes, incorrect feedback signal from the color oscillator due to some component failure, or a defect in the phase shift amplifier. This latter defect is sometimes due to shorted capacitors which cause increased amplifier gain and therefore some instability. Whatever defect is finally found, the entire color synch section must usually be adjusted for best performance. The procedure presented in Chapter 15 should be followed closely.

In the preceding paragraphs the 7 most frequent color defects were discussed. Occasionally symptoms may be found which do not seem to fit into any of these categories; these may stump the serviceman temporarily. A brief consideration of the following questions will help to classify such a defect and fit it into one of the categories listed, either in this chapter or in Chapters 17 or 19.

(a) Which colors are missing?
(b) Which colors dominate?
(c) What does a white object look like?
(d) How does the picture look without color?
(e) Which type color decoder is used?
(f) How does variation in brightness or contrast affect the coloring?

Questions (a) (b) and (c) help to identify the section or primary color responsible. Question (d) relates the defect to the Y signal; the last two questions help determine the troubleshooting procedure.

19

MISCELLANEOUS
TROUBLESHOOTING

Chapters 16, 17, and 18 have explained the defects peculiar to color TV receivers in regard to monochrome and color operation. Those defects which are not due to component failure but are caused by interference or misalignment of the receiver were omitted till now. In this latter category we will find such instances as sets with which the customer himself has tampered as well as receivers which either were skipped by the manufacturer's test department or the happily few cases where another serviceman has already tried his luck. Alignment procedures for each receiver section are presented in Chapters 13, 14, and 15. In this chapter we will cover those defects which are either caused by completely wrong alignment or can be corrected by special alignment, or those caused by some kind of interference—internal or external. Toward the end of this chapter we also present a method for tracking down intermittent defects and finally give some hints on troubleshooting those "impossible to find" defects that are the bane of the service technician.

Interference is not confined to television; it causes trouble in other fields of communication as well. The same types of interference that mar radio and monochrome TV reception will also wreck the color picture. Remedies applicable to monochrome TV are usually also useful for color, except in those instances in which the remedy requires limiting the RF or IF bandwidth.

Defects of the type listed in this chapter are generally rare and for that reason the average serviceman will not be too familiar with them. Changing tubes, measuring voltages, and tracing video signals are not sufficient to troubleshoot these defects and the student may sometimes want to give up and admit that he is stumped. Actually, each of these defects can be located by applying a few tests and then, based on a good knowledge of the receiver operation, the trouble spot can be reasoned out. We have purposely left a discussion of these defects for the end of the book since it is hoped that the reader will have gained at this point an understanding of the color receiver that enables him to figure out even the most baffling defects.

Interference can appear in both sound and picture or only in one of the two. The interference can originate outside the set or it can be generated by the set. Each type of interference can be isolated; while there are some instances in which the only positive remedy would be to dynamite the interfering oscillator next door, most interference troubles can be corrected.

External Interference

To ascertain that the interfering signal orignates outside the receiver the following steps can be followed:

1. Short antenna terminals and see if the interference is still there.
2. Is the interference observed on all channels? At all hours? On monochrome as well as color? These questions will help determine the exact type of interference. Amateur transmitters cause interference which may be present even without antenna and appear on all channels, but the interference lasts only for a few minutes at a time, then goes off and then appears again. Furthermore, the amateur usually does not operate during the daytime.

Diathermy interference appears locked in with the picture because of the 60 cycle common power supply. This interference may last for 15 minutes at a stretch but usually goes off after the doctor's office hours.

Police radio interference will appear all the time, but each time only for a few minutes. It may appear on all channels. Neighboring aircraft beacons, radar installations, or similar transmitters will cause continuous interference, usually confined only to one channel.

Interference due to ignition from passing vehicles appears as streaks racing across the screen. To eliminate this interference it may be necessary to move the antenna and transmission line away from the street side of the house, use shielded cable and try various filters.

FROM (mc.)	TO (mc.)	SERVICE	HARMONIC	ENTERS AT	REMEDY
.535	1.605	BC	—	2nd det., video, a.c.	Shielding, line filter
1.605	1.8	Aircraft, police	—	2nd det., video, a.c.	Shielding, line filter
1.8	2	Amateur, loran	—	2nd det., video, a.c.	Shielding, line filter
2.065	2.105	Maritime	10	I.F.	High-pass filter, shielding
3.2	3.4	Short-wave	7-8 / —	I.F. / 2nd det., video, a.c.	High-pass filter / Shielding, line filter
3.5	4	Amateur	6-7 / —	I.F. / 2nd det., video, a.c.	High-pass filter / Shielding, line filter
4.75	4.85	Short-wave	5	I.F.	High-pass filter, i.f. shields
7	7.3	Amateur	3	I.F., antenna	High-pass filter, i.f. shields
11	11.975	Short-wave	2	I.F., antenna	High-pass filter, i.f. shields
14	14.35	Amateur	3	I.F., 41 mc., antenna	High-pass filter, i.f. shields
			4	Antenna, Channel 2	Modify antenna and lead-in
15.1	15.45	Short-wave	3	I.F., 41 mc., antenna	High-pass filter, i.f. shields
17.7	17.9	Short-wave	3	Antenna, Channel 2	Modify antenna and lead-in
21	21.45	Amateur	—	I.F., antenna, a.c.	High-pass filter, shielding, line filter
21.45	21.75	Short-wave	—	I.F., antenna	High-pass filter, i.f. shields
21.85	22	Aircraft	—	Antenna	High-pass filter
22	22.72	Maritime	—	I.F. (rarely), antenna	High-pass filter
23.2	23.35	Aircraft	—	Antenna	High-pass filter
25.6	26.1	Short-wave	—	I.F., antenna	High-pass filter, i.f. shields
27.185	27.455	Industrial, medical	2	Antenna, Channel 2	Modify antenna
			—	I.F., antenna, a.c.	High-pass filter, shielding, line filter
28	29.7	Amateur	2	Antenna, Channel 2	Modify antenna and lead-in
29.7	42	Gov't, fire, petroleum, etc.	2	Antenna, Channels 2-6	Modify antenna and lead-in
			—	I.F., 41 mc.	High-pass filter, shielding
42	44	Police, highway	—	I.F., 41 mc., antenna, a.c.	High-pass filter, shielding, line filter
44	50	Short-wave	—	I.F., 41 mc., antenna, a.c.	High-pass filter, shielding, line filter
54	72	TV	Osc. radiation	Antenna	Modify antenna and lead-in, re-align and shield offender
76	88	TV	Osc. radiation	Antenna	Modify antenna and lead-in, re-align and shield offender
88	108	FM	2	Antenna, Channels 7-13	Modify antenna and lead-in
			Image	Antenna, Channels 2-3	Wave trap
108	132	Aircraft	Image	Antenna, Channels 3-6	Wave trap
148	174	Gov't, railroad, police	Image	Antenna, Channels 7-13*	Wave trap
			3	Antenna, Channels 14-22	Modify antenna and lead-in
174	216	TV	Osc. radiation	Antenna	Modify antenna and lead-in, re-align and shield offender
216	220	Gov't	3	Antenna, Channels 43-46	Modify antenna and lead-in
220	225	Amateur	Image	Antenna, Channel 7	Wave trap
			3	Antenna, Channels 46-49	Modify antenna and lead in
225	470	Aircraft, military	Image	Antenna, Channels 14-28*	
			2	Antenna, Channels 14-83	Modify antenna and lead-in
470	890	U.h.f. TV	Osc. radiation	Antenna, Channels 14-83	Modify antenna and lead-in, re-align and shield offender

*Where oscillator operates below incoming signal.

Figure 19.1—TABLE OF INTERFERENCE SOURCES

An interference particularly difficult to overcome is the one caused by radiation from the local RF oscillator in a neighboring TV set. Usually this interference will be observed only at one or two channels and only at certain times. If both the customer and his neighbors have only one station to watch, this type of interference is not likely.

There are many other types of interference caused by external signals. A list of frequencies associated with the particular service causing TV interference is shown in the table of Figure 19.1 together with a suggested means for eliminating it. AC power line filters can be purchased and simply plugged in between the power line and the TV receivers. Other remedies listed in Figure 19.1 depend on the point at which the interfering signal enters the receiver. Whenever possible a filter in series with the antenna lead-in line should be installed, making sure that the filter does not wreck the bandpass characteristic. A simple method of checking filters is to connect the filter in the circuit shown in Figure 19.2a, together with a sweep generator and oscilloscope. Be sure to use a satisfactory detector of the same impedance as the filter, usually 300 ohms. Such a detector circuit is shown in Figure 19.2b. When the sweep generator is tuned to each of the channels

Figure 19.2—(a) SWEEP METHOD FOR FILTER TESTING; (b) 300 OHM BALANCED DETECTOR CIRCUIT

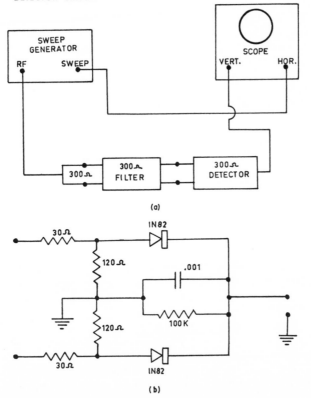

being received the oscilloscope should show the same flat response curve as if no filter were used at all.

Wavetrap stubs consisting of a quarter wavelength of transmission line are not broadband but, if tuned to an adjacent channel, such traps can be used for color TV. In general the length of such a tuned stub can be approximately calculated by the equation given below; exact tuning is done by cutting the transmission line in ¼-inch steps and observing the effect.

$$\text{Length (inches)} = \frac{2950}{\text{Frequency (mHz)}}$$

In some instances the interference cannot be eliminated by traps or shielding of the transmission line, but the entire antenna and line must be moved to a different location. Such modifications in the original installation are fairly costly and should only be done if a temporary alternate installation proves to the customer that a clean signal can be obtained.

The too-frequent case in which a neighboring TV receiver interferes without the offender being aware of it is best handled by offering to shield and decouple the offending oscillator. In some instances the introduction of a booster between the radiating tuner and the antenna is sufficient to cure the trouble.

In a few cases the interference may be so strong that it is picked up by the chassis itself or some of the components. The final remedy in such a predicament is to shield the entire receiver with copper screening inside the cabinet.

Internal Interference—Oscillation

Whether the set is operating on monochrome or color, oscillation in the IF or RF section can cause complete obliteration of the video signal or, if the oscillation is not too strong or the set only slightly regenerative, the picture may appear streaked. This condition may only occur on weak stations where the AGC bias is low and does not reduce the IF gain enough to stop the regeneration. Adjust the bias control, if available, to check for this type of defect.

Regeneration is most frequent in the IF and RF sections, but also can occur in the sound or video amplifiers. The troubleshooting method outlined below will help in locating the stage in which the regeneration takes place and then the various bypass and decoupling capacitors should be checked to pinpoint the guilty component.

1.   Short antenna terminals. This should not affect the interference pattern unless it is caused by an outside signal and not regeneration.

2.   Disable the local oscillator in the tuner. If the oscillation stops, it originated in the tuner. If it continues, the regeneration takes place in a subsequent stage.

3.   Disable the last IF amplifier or the video detector tube. Should the interference stop, the IF section is at fault.

4.   Disable the second video brightness amplifier. If the trouble is in the brightness channel, the interference will stop.

5.   Disable the color synch oscillator or, in receivers using a crystal ringing circuit, the last limiter amplifier. This will serve to cut off the color demodulators and, if interference persists, it must be in some stage between the demodulators and the color picture tube.

6.   Disable each chroma amplifier in turn to locate any regeneration in that section.

7.   Disable the audio IF and amplifier to eliminate them as a possible source.

8.   As a final check, try shunting a .05 mfd capacitor across various points in the B+ supply to make certain that it does not contribute to the instability. Next, bypass the AGC bias points in a similar manner to insure that the filtering is correct.

In addition to defective components, the poor initial alignment and aging of tubes, transistors, etc. occasionally cause regeneration due to mistuning of one or more coils. A complete alignment process may be required to properly cure regeneration.

Internal Interference—Beats

A typical beat pattern that might be due to misalignment of the sound IF trap is shown in Figure 19.3. Here, the sound is a 1 kHz tone.

Figure 19.3—SOUND BEAT PATTERN

This defect appears the same in both black and white and color TV. In the color receiver there is more likelihood of sound beats because the video bandwidth has to be extended to within 500 kHz of the sound carrier. To complicate matters, the color sub-carrier is about 900 kHz from the sound carrier so that a 900 kHz beat signal can become troublesome in the color TV receiver. Figure 19.4 shows this type of defect. Other beat signals that will mar reception can occur between the received signal and those from adjacent channel sound or video carriers, or even from adjacent channel color sub-carriers. These external beats can be determined by shorting the antenna terminals. If the beat pattern disappears, it is due to some picked-up signal. One remedy is to use a narrow beam antenna and orient it for reception only from the desired station. Antenna rotators are often used in such a situation to allow reception on different channels without interference. Be sure, that all adjacent channel IF traps are properly aligned, before changing the antenna.

When it is established that the beat signal is due to an internal defect, the sound carrier should be suspected first. One indication of this is the fact that by misadjusting the fine tuning control on the tuner, the beat can be reduced or eliminated entirely. Once this test has been made, the chassis must generally be removed from the cabinet to allow detailed trap alignment checks. A complete alignment procedure for the IF section and all traps has been presented in Chapter 13 and at first only the trap adjustment need be done. If adjusting the sound IF traps does not correct the beat pattern, then a complete overall alignment is indicated.

Occasionally beats result from excessive coupling between the audio and video sections. Such coupling may be due to open decoupling capacitors, shorted chokes, or even displaced wiring moved during the repair of some other defect. Be sure all ground wires are soldered well and all shield cans are firmly grounded. Missing tube shields or floating grounds on shielded cable can introduce a beat note. If the beat appears as a fine grain pattern only on color pictures, then the most likely offender is the 900 kHz interference caused by the color sub-carrier and sound IF carrier beat. A coarser pattern which appears to move with the sound itself indicates slope detection of the audio in the video detector and can usually be observed on color as well as on monochrome transmission. Proper sound IF trap alignment is again the usual solution.

In some receivers there is either no automatic color killer circuit, or that section might not operate correctly or may not be able to cut-off the color oscillator. When the 3.58 mHz oscillator is on during a monochrome telecast, a fine beat signal can appear in the picture, shifting and weaving but always distinguishable by the fact that it is fine grain. When a color-black and white switch is used, be sure this switch is in the correct position; where a color killer circuit is employed, its performance should be suspected. Just to make certain that the color synch signal is the source of the interference, disable the 3.58 mHz oscillator. This should eliminate the interference. To troubleshoot a defect of this type it is invariably necessary to remove the chassis from the cabinet and check various voltages.

First, check the color killer circuit voltages. Next, look for poor grounds in shielding or decoupling circuits which normally keep the 3.58 mHz signal from the screen. Finally, check the alignment and operation of each of the 3.58 mHz traps in the video and color sections. Defective coils or capacitors in these traps may cause the trouble, or some mechanical defect—floating grounds, ungrounded shielding, or the like—could be at fault. Tracing this type of interference requires patience and experience as well as an understanding of the circuitry.

As mentioned before, all of the troubles causing interference pattern in monochrome receivers can also appear in color sets; this includes such familiar plagues as arcing, Barkhausen oscillations, 60 or 120 Hz hum, and so forth. The same remedies as in monochrome servicing will also apply in color TV. Listed below are some of the symptoms and remedies for these internal interference troubles to serve as ready reference.

Thin vertical bars along one edge of the raster. This is usually due to Barkhausen oscillations, a defect originating in the horizontal output amplifier. Try changing tubes and rearrange wiring to deflection yoke and flyback transformer. Finally, try locating a small permanent magnet somewhere on the envelope of the horizontal output amplifier tube until the interference disappears.

Arcing. This looks like ignition interference, but is irregular. It may be apparent only on damp days, with high line votage conditions. Try to see the arc-over point directly in the HV supply. Removing sharp corners, adding HV insulation and possibly coating the offending part with a good insulating liquid will help.

60 Hz hum. This may appear as a dark horizontal bar extending over the entire screen, if the interference is in the video signal. If it is in the vertical sweep, there will be poor hold and jitter. If 60 Hz gets into the horizontal sweep circuit, scalloped edges and weaving of the picture will occur, as in Figure 19.4. If the colors seem to vary in strips or get lighter and darker in a weaving motion, the 60 Hz interference is most likely in the color synchronizing or phasing circuit. The usual cause of 60 Hz interference is heater-to-cathode leakage in some tube since the power supply generally uses a full wave rectifier and its interference would then be at 120 Hz.

120 Hz hum. This may take the same form as the 60 Hz hum, but can be distinguished if it can be locked in with the picture temporarily. In the case of hum in the video section, there will be two horizontal bars, either dark or light for 120 Hz interference as shown in Figure 19.5. The 60 Hz hum will have half the number of bars. Remedies for 120 Hz hum include replacement of some B+ filter capacitor or decoupling network.

Microphonics. Tapping the set causes visible jitter or even complete picture breakup. This is invariably due to either an intermittent connection or an intermittent component. Most frequently tuner contacts, tube pins or the pigtails on resistors, coils or capacitors are broken and cause microphonics. Bad grounds, tubes partly out of the socket, and so forth can also be responsible. One difficulty arises when the microphonics are continuous due to some constant jarring such as vibration due to some appliance. The customer may then think that he is plagued by some outside interference such as ignition arcing while actually the set is microphonic.

Ghosts or reflections. One of the most annoying troubles in TV reception is the appearance of multiple images [Ghosts]. These ghosts are usually due to reflections from tall buildings or mountains, etc., which causes two signals, one slightly delayed from the other, to appear at the TV receiver antenna. In many metropolitan areas ghosts on monochrome reception are so frequent that the viewers have come to accept them, but in color TV ghosts are much more annoying because they result in apparent color misregistration. Ghosts on color pictures are often mistaken for poor convergence and vice versa. The first object of the technician in troubleshooting this type of defect is therefore to distinguish between poor convergence and a ghost.

Tune in a monochrome transmission or else reduce the color gain control so that a monochrome picture appears. If poor convergence is the cause, the edges in the monochrome picture will appear in two or three colors, like the poor convergence illustrations of Figure 15.10 and 15.11. The appearance of a ghost does not lead to any color fringing in monochrome. Figure 19.6 shows a typical full color ghost and how it appears on monochrome is shown in Figure 19.7. In the full color picture of Figure 19.6 portions of the flesh color are transposed to the right, and portions of the blue background are transposed to the left portion of the face. The colors appear smeared to the right. In the monochrome picture only a slight amount of shift to the right can be seen and this is hardly noticable and certainly not as objectionable as in color. To demonstrate to the set owner just what the problem is, it is often a good idea to adjust the set to a single color, such as the illustration of Figure 19.8. Here the outline of the ghost, shifted to the right, is clearly visible and its disturbing effect on the total color picture can be seen.

This example also illustrates that a ghost which may be barely noticable on monochrome transmission becomes extremely disturbing on color. For this reason the elimination of ghosts often plays an important part in providing good color reception.

Color TV ghosts can originate in two ways. The most frequent ghost is due to reflections between the transmitter and the receiving antenna. To eliminate this type of ghost it is necessary that an antenna location, direction and pattern be found which will eliminate the unwanted reflection or at least attenuate it greatly. The usual solution is to select a very directive antenna and to try and orient it to the very best possible picture. In some locations it is impossible to get a ghost free signal and the set owner should be made to understand this.

Reflections need not always be due to obstacles or reflectors in the signal path between the transmitter and receiver but can also occur due to impedance mismatch between the antenna and the receiver terminals. Such a mismatch will cause some of the signals to be reflected back and forth between the antenna and the receiver terminals. To make sure that the transmission line or any auxiliary devices such as multi-set couplers or preamplifiers are not causing the ghosts, they can be checked for "Standing Waves." On an unshielded transmission line wrap a 10″ piece of aluminum foil around it, grasp it by the hand and slide it up and down over a length equivalent to at least one quarter wave length at the channel at which the ghost occurs. If any change in the ghost is noted during that procedure, it is probably caused by the transmission line or the devices connected to it. Transmission line ghosts can be reduced by terminating the transmission line with its characteristic impedance. In flat ribbon lines this is usually 300 ohms and a simple 300 ohm resistor inserted at the right spot in the transmission line may go a long way toward reducing the effect of this type of ghost.

In many installations, particularly in strong signal areas, the transmission line itself acts as receiving antenna. Even though the antenna is directed to receive only the desired signal, some of the reflections may be picked up by the transmission line. In such cases a shielded transmission line, either of the two-wire balanced type or the coaxial type is recommended.

A host of antennas, transmission lines and accessories are now available for color TV installations, all designed to minimize the problem of reflections or impedance mismatch. In selecting the right one it is important to consider not only the frequency response but the phase response or, as it is usually expressed, the voltage standing wave ratio [V.S.W.R.]. A good VSWR will be approximately 1.15:1 for the input and 1.4:1 for the output, absolute maximum. This means that the ratio of reflections to the desired signals will be quite low as concerns the transmission line and the antenna coupler or amplifier.

In Chapter 3 we have described the phase relationships of the color synchronizing signal and the chroma sub-carrier and have seen the importance of maintaining this phase relationship. If the tuner, I.F. and other wideband circuits are properly aligned, the phase delay of the color sub-carrier or the reference burst will be uniform as it passes from the antenna through the receiver. It is, however, possible that certain defects or poor alignment can cause a non-linear phase delay in the receiver. Figure 19.9 shows the RF response of channel 4 of a color TV receiver and the phase delay of the same frequency band. If the phase delay varies by more than $\pm 5°$ over the frequency band of the color sub-carrier this can cause phase distortion, appearing as wrong colors.

Remember that the individual colors, as they are demodulated, are a result of the phase difference between the reference or color synch signals and the color sub-carrier. If the color sub-carrier phase is wrong due to uneven phase delay in the receiving section, this will appear as wrong colors. For example, to produce a given shade of magenta a certain amplitude of $-X$ and $-Z$ vectors are decoded. If the phase of the X and Z vectors are shifted with respect to the reference burst, the resulting colors will also be different. If the X vector is shifted by 10° while the Z vector is shifted by 15° the result will be an entirely different shade, more of a purplish blue.

To differentiate between a ghost caused by signal path problems and one caused by uneven phase delay in the receiver, a color bar generator can be connected to the antenna terminals and the resultant color bar pattern can be inspected. If the color bar pattern appears correct then the ghost is due to external reflections. If unequal phase delay is the source of the trouble then one or more of the color bars will appear in the wrong color.

The remedy for unequal phase delay in the receiver is usually a realignment of the circuits causing this problem. As was pointed out in Chapters 8 and 13, the flatness of the frequency response of both the tuner and the IF section is essential in keeping phase delay variations to the minimum of less than 5%.

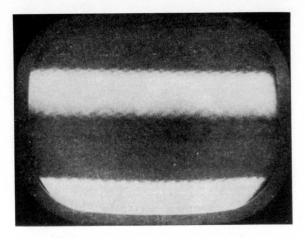

Another type of ghost occurs in very rare occasions when the delay line in the brightness channel becomes partially defective. If this component is not properly terminated, the brightness signal is reflected back and forth from the ends of the delay lines, causing several ghosts, evenly spaced across the screen. When a color bar generator is used, the same ghost pattern will result and in that case the delay line is invariably the suspected defective part.

Locating Intermittents

Every technician knows how hard the elusive intermittent defect is to troubleshoot. One method of servicing intermittent defects endeavors to make the intermittent defect permanent and then find the trouble by the usual method. This procedure employs three factors which most frequently will show a real defect up. They are: vibration, voltage, and atmospheric conditions. The procedure described below will help to smoke out most hidden and intermittent defects.

1. Observe the intermittent defect at least once to get some idea of the receiver section at which it originates. For example, if the horizontal hold is lost intermittently, the defect is most likely in the horizontal AFC section. If the picture disappears, but sound continues, the HV may be at fault.

2. With the set operating correctly, tap each tube, part, cable, and shield with a rubber mallet. Tap and check the various grounds and shielding points. If this does not locate the defect, proceed to the next step.

3. Connect the set to a variable AC line voltage supply such as a Variac. Adjust the line voltage to 95 and wait for about 10 minutes for the defect to reappear. Next, set the line voltage to 125 volts and

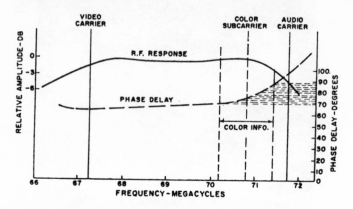

Figure 19.9—RF RESPONSE AND PHASE DELAY

let the set operate for an hour or more. Frequently the excessively high voltage will speed up breakdowns in any intermittent component.

4. In a stubborn case it may be necessary to use high voltage as well as temperature to smoke out the guilty component. One simple yet effective way to accomplish that is to close the ventilating louvers and rear cover with a blanket which can be thrown over the entire receiver. Leave the line voltage at 125 and allow the set to run for several hours in this manner. Usually the intermittent part will break down permanently within less than half an hour. To make the heating more effective, especially in the case of suspected HV trouble, a small steam vaporizer can be put under the blanket to raise the humidity.

It may be thought that this type of treatment is abusive and might damage the TV set, but actually the temperature and humidity conditions in some of the southern seashore communities are often just as severe. Furthermore, each component in the receiver is rated for more than the heat, humidity, and voltage which it normally will encounter. Therefore good parts will be able to withstand this test while the partially defective items will become completely bad in a very short time. Since many of the intermittent defects are due to parts just on the verge of becoming permanently bad, the method outlined above will work in more than 90 per cent of all instances.

Troubleshooting the "Impossible" Defects

When a TV set has gone through a good check at the customer's home and has been tested in the service shop without yielding some clue as to the defective part, many servicemen either give up and return the set or else try to rebuild portions of it in the blind hope that somehow they will inadvertently repair it. Neither course should be necessary for the skilled service man. The important thing is to keep a clear head and not be tempted by wild schemes and half-baked conclusions. There is no need for despair or panic since, "if a man made it, a man can fix it" is as true for TV receivers as for any other mechanical device. It is not feasible here to give a detailed procedure for finding every conceivable trouble, but some suggestions are presented which will help with difficult defects.

Often, after working on a set for a long time, things become confused, wrong tubes are used as replacement, connections are left open, cables partly connected, shorts left in, and so on. To keep confusion from overcomplicating a job, be sure it has been determined that the set is going to be a problem child. After this is done, preferably by a second technician, the best thing would be to postpone further work on the set and tackle something else. Many times a short break for coffee, lunch, or sleeping on a troubleshooting problem have led to a simple solution, where some obvious tests or faults have been overlooked.

When the problem child is attacked again, forget any previously used procedure and follow the steps listed below, writing down any data taken.

1. Measure all voltages according to manufacturer's voltage chart or schematic diagram data. Record them.

2. Perform complete RF, IF, sound, and color alignment. Be sure to write down all frequencies and the tuned networks which correspond to them. Note any points which seem hard to align.

3. Trace the video and color signals from the detector to the kinescope with an oscilloscope noting the amplitude changes occurring in each stage.

4. Repeat step 3 for vertical, horizontal, and color synch sections. Trace the sweep sections through.

5. Go over the notes taken and compare the data with the manufacturer's data or with those from a similar receiver.

If there is still no definite clue to the defect the substitution method can be used as last resort. A color TV set of the same model or a similar one is placed alongside the problem child. Step by step, each section of the good set is connected in place of the corresponding section of the bad set. Be sure that both sets have transformers in the power supply or else use an isolation transformer to avoid AC shorts. Both sets should be connected to the same antenna and the same channel. Although this method is cumbersome and requires the presence of a second good set, it will invariably turn up the most stubbornly hidden defect.

As was stated before, any and every defect can be repaired. Lest the reader be misled into thinking that he will encounter great difficulties in every service job, some statistics of a typical servicing organization specializing in color television might be quoted here. Eighty per cent of all service calls involved replacement of tubes or capacitors. Forty-five per cent of all calls were completed in the customer's home, in an average time of 20 minutes per call. Of the 20 per cent of defective color receivers which required parts other than tubes or capacitors, 18 per cent were repaired in less than 1 hour of actual bench time. Only 2 per cent of all sets required the attention of more than one service technician and needed thorough troubleshooting. This 2 per cent included a number of sets on which service had been refused by another service technician, sets which had fallen off the delivery truck, receivers which had been "fixed" by the owner, and so on.

With these statistics in mind the reader can retain his confidence in the feasibility of servicing color TV receivers. Only a good understanding of the principles of color TV, a knowledge of receiver circuits and their operation, and a clear head are needed to be a successful TV serviceman, in monochrome as well as in color TV.

INDEX